ELECTRICALLY CONDUCTIVE ORGANIC POLYMERS
FOR ADVANCED APPLICATIONS

ELECTRICALLY CONDUCTIVE ORGANIC POLYMERS FOR ADVANCED APPLICATIONS

by

David B. Cotts and Zoila Reyes

SRI International
Menlo Park, California

NOYES DATA CORPORATION
Park Ridge, New Jersey, U.S.A.
1986

Copyright © 1986 by Noyes Data Corporation
Library of Congress Catalog Card Number 86-18053
ISBN: 0-8155-1094-2
Printed in the United States

Published in the United States of America by
Noyes Data Corporation
Mill Road, Park Ridge, New Jersey 07656

10 9 8 7 6 5 4 3 2 1

Library of Congress Cataloging-in-Publication Data

Cotts, David B.
    Electrically conductive organic polymers for
advanced applications.

    Bibliography: p.
    Includes index.
    1. Polymers and polymerization--Electric properties.
I. Reyes, Zoila. II. Title.
QD381.9.E38C68   1986   620.1'9204297   86-18053
ISBN 0-8155-1094-2

Foreword

This book is a study of electrically conductive organic polymers for advanced applications. The properties of electrically conducting, semiconducting, and semi-insulating polymers were surveyed and their conduction mechanism, mechanical properties, and suitability for space-based use evaluated. Correlations between molecular structure, conductivity, and mechanical properties were drawn, and a comprehensive model of electrical conductivity in organic polymers was formulated.

The environmental exposure of polymer dielectrics used in spacecraft to high-energy electrons results in the accumulation of secondary electrons. When the electrostatic potential resulting from the accumulated charge exceeds the dielectric strength of the polymer, a breakdown occurs that can interrupt or damage normal function. This charging phenomena could be eliminated if dielectrics with a moderate electrical conductivity were used. At high potentials, charge would be drawn off at a higher rate than electrons are accumulated, thus preventing discharge. A successful material for the moderation of charging problems would thus possess a moderate resistivity and the other properties required for space-based use, including mechanical integrity, environmental resistance, and long-term stability.

Although the most widely studied electrically conducting polymers are not robust enough for most space-based uses, several commercial materials appear to have the necessary combination of electrical, thermal, and mechanical properties. The main obstacle to the selection of new or modified materials for spacecraft use is the lack of strength, thermal stability, and radiation resistance—not their conductivity. Several synthesis procedures are identified that would raise the value of these properties to acceptable levels for materials that have the required electrical properties.

The wide range of data reported in the literature can be reconciled by a theory of conductivity in which the limiting feature is the rate at which electrons are transferred between localized charge states. Variations in chemical structure lead to changes in the relative energy of the charge states and their relative orientation, but high mobilities are observed only in systems that form a periodic superlattice, of localized states. Some of the new polymers identified by this model have been prepared. They possess relatively high electrical conductivities and, unlike the majority of electrically conducting polymers, are processable in organic solvents. Although prepared for space applications, this book has obvious commercial implications. The chemical and polymer industries should benefit considerably from this book, as it places between covers the appropriate theory and technology, as well as brings together data regarding the electrical conductivity of about 250 polymers. The book is a very good guide to current technology, and future applications.

The information in the book is from *New Polymeric Materials Expected to Have Superior Properties for Space-Based Uses,* prepared by David B. Cotts and Zoila Reyes of SRI International for Rome Air Development Center Air Force Systems Command, July 1985.

The table of contents is organized in such a way as to serve as a subject index and provides easy access to the information contained in the book.

ACKNOWLEDGMENTS

A project of this scope invariably requires the assistance of many individuals. In particular, the authors gladly acknowledge Carolyn Kelly, John Rose, and Amy McNeil who were largely responsible for organizing the literature search, performing laboratory experiments, and preparing our reports, respectively. Valuable suggestions and background information were provided by Dr. A.R. Frederickson, the USAF Project Monitor. Finally, we thank the SRI library staff who assisted in locating obscure journal articles and patiently reshelved countless journals and texts.

NOTICE

The materials in this book were prepared as accounts of work sponsored by the U.S. Air Force Rome Air Development Center. Publication does not signify that the contents necessarily reflect the views and policies of the contracting agency or the publisher, nor does mention of trade names or commercial products constitute endorsement or recommendation for use. To the best of our knowledge the information contained in this publication is accurate; however, the publisher does not assume any liability for errors or any consequences arising from the use of the information contained herein.

Contents and Subject Index

1. Introduction

The environmental exposure of polymer dielectrics used in spacecraft to high energy electrons results in the accumulation of secondary electrons. When the electrostatic potential resulting from the accumulated charge exceeds the dielectric strength of the polymer, a breakdown occurs that can interrupt or damage normal spacecraft function. This spacecraft charging phenomena could be eliminated if dielectrics with a moderate electrical conductivity were used. At high potentials, charge would be drawn off at a higher rate than electrons are accumulated, thus preventing discharge.

A successful material for the moderation of spacecraft charging problems would thus possess a moderate resistivity, assumed to be about 10^{12} ohm cm, and the other properties required for space-based use, including mechanical integrity, environmental resistance, and long-term stability.

The purpose of this report is to formulate a coherent theory of conduction in polymers to allow the identification of materials that have resistivities less then 10^{12} ohm cm and that meet other requirements for spacecraft use. The report structure is as follows: In Section 2 we summarize our results and conclusions. The following two sections (3 and 4) deal with the contemporary use of polymers in spacecraft and discuss the effects of the radiation found in the spacecraft environment on the properties of polymers. The known conducting and semiconducting polymers are discussed in Section 5 and their suitability for use in space is discussed in Section 6. The next two sections summarize contemporary models of the conduction process (Section 7) and present a generalized model for conduction in organic polymers (Section 8). The final two sections (9 and 10) discuss the implications of this model on the design of new or modified electrically conducting polymers and the important questions concerning polymer conductivity that remain to be answered by future research.

Each of these sections is largely self-contained. We hope that, for example, an engineer can use the results of our material evaluation in Section 6 without becoming embroiled in the polymer structures presented in Section 5 or the discussion of polymer conductivity in Section 7. Similarly, we hope that the chemical community will benefit from our conclusions concerning structure-property relationships in Chapter 9 without necessarily digesting the information about the applications of polymers discussed in Section 3. For convenience, our results and conclusions are summarized in the following section.

2. Summary and Conclusions

Despite the extensive range of polymer compositions and structures that have been investigated over the last 50 years, there is no good theoretical model for conductivity or empirical correlation between polymer structure and electrical conductivity. One reason for this is that polymers possess a heterogeneous degree of inter- and intramolecular order. Good models exist for crystalline and amorphous semiconductors which represent two extremes in structural order. However, no good model exists for materials with intermediate degrees of order.

Various polymer conductivity phenomena can be explained by introducing the inhomogeneous transport of charge through ordered regions with periodic differences in composition. These disparate phenomena include the dramatic effect of oxidizing agents or dopants on conductivity, mobilities ranging over 10 orders of magnitude, a metallic thermopower, and a temperature dependence of the conductivity indicative of a variable-range hopping mechanism.

In spacecraft applications a relatively small set of commercial materials are used to fulfill a specific set of functional requirements. Generally, the relationships between molecular structure and material property (e.g., thermal stability, radiation resistance, strength) are better known than the dependence of conductivity on structure, so there is considerable lattitude in which to design new materials or to modify known polymers. To identify these modifications, we have brought together data reported in the literature concerning the molecular structure and electrical conductivity of nearly 250 polymers. These data indicate that inhomogeneous transport between localized states is responsible for the temperature, dopant, processing, and structure dependence of conductivity in organic polymers. The structural features favoring conductivity include:

3

- A polymer backbone containing medium-range (10-50 $\overset{\circ}{A}$) electron delocalization.

- Molecular orbital structures capable of stabilizing radical ions, particularly heteroaromatic rings containing nitrogen and sulfur.

- Few or no pendant groups to interfere with close intermolecular packing and alignment.

- Ability to form partially oxidized (or reduced) complexes with intercalated low molecular weight species.

On the basis of these general descriptions, several contemporary or readily modified materials have been identified that we believe have the requisite conductivity to reduce spacecraft charging and the physical properties necessary for application in spacecraft.

One example is polyvinylcarbazole (PVK), which has good thermal stability, radiation resistance, mechanical properties, and is a good semiconductor. It has been extensively investigated as a photoconductor and can be doped with additives or its molecular structure can be altered to control its conducting properties. Development of more highly stereoregular PVK could increase its thermal stability by 50° to 100°C, making it useful as a primary structural material for direct exposure to the space environment.

A second potential class of materials are the pyropolymers including pyrolyzed polyacrylonitrile (PAN), pyrolyzed aromatic polyimides (e.g., Kapton), and polyacene quinone radicals (PAQRs). The electrical properties of pyrolyzed PAN and Kapton have been studied in some detail. Although their molecular structure after pyrolysis is relatively unknown and very little mechanical property data are available in the literature, they appear to be of considerable potential in reducing spacecraft charging phenomena. Carbonized PAN fibers are the primary source of high-strength carbon fiber, which is widely used in the aerospace community, so it is reasonable to believe that a pyrolyzed PAN or polyimide would have significant strength and stability. Kapton film and other polyimide resins are already used in aerospace vehicles and valued for their strength, radiation resistance, and thermal stability.

A third possibility is the use of modifications of the four most highly conducting polymers: polythiazyl (SN_x), polypyrrole (PP), polyphenylene sulfide (PPS), and polyacetylene (PA). The high conductivities of these materials in their doped "metallic" state are indicative of an electron delocalization that is incompatible with chemical or thermal stability. Undoped materials display nearly semiconducting (10^{10}–10^{12} ohm cm) resistivities that are believed to be adequate for protection from spacecraft charging. Very little mechanical property data are available for these materials, and more work needs to be done if they are to be qualified for spacecraft use.

Finally, there are new polymer compositions that meet the four criteria summarized on the previous page. We explored the synthesis of these compounds and have found several that have the desired electrical properties in addition to being mechanically strong, stable, and easily processed into films, fibers, and castings. Many of these polymers are structurally similar to the more radiation-resistant polymers and may represent a new class of materials for use in minimizing the problem of spacecraft charging.

This study has uncovered many questions concerning conductivity in organic polymers, its relation to molecular structure, and its application to problems as diverse as spacecraft charging, and molecular electronic devices. The basic methodology used (i.e., correlating molecular structure with mechanical, thermal, and electrical property data) has provided guidelines for the selection of spacecraft materials and insight into the mechanisms responsible for polymer conductivity

3. Polymer Use in Spacecraft

This section consists of three parts. The first describes the general applications in which polymers are used in spacecraft to illustrate the diversity of materials and properties encountered. The second contains a set of generic material requirements for each of the applications discussed. Although these descriptions may not be exact for each possible example, they provide a set of baseline properties for comparison with new materials. The third section summarizes the types of commercial materials used in these applications. In reading material specifications, one is struck by the fact that although many formulations may contain a readily available polymer (e.g., polystyrene), few materials with alternative molecular structures and comparable thermal and mechanical properties have been evaluated.

Our objective in preparing this section was to provide the reader with a functional description of specific applications, an example of the materials used, and a sense of the range of molecular structures that could be used. The following text is an introduction to the bulk of the data in Appendices A through E.

Functional Applications of Polymers

A series-by-series (e.g., Explorer, Pioneer, Nimbus) description of the polymeric materials used in spacecraft can be found in the Space Materials Handbook,[1] which illustrates five general categories of applications:

- Structural polymers including films, casting resins, fibers, and matrix resins.

- Electronic applications including insulation, circuit boards encapsulants, and standoffs.

- Adhesives, sealants, and elastomers.

- Lubricants, both solid and liquid.

- Thermal control including insulation and radiative surface coatings.

Material Requirements

The use of organic polymers in structural applications is increasing rapidly because of significant weight savings. Concomitant advances in the development of thermoset and thermoplastic resins with improved high temperature/environmental resistance are just beginning to have an effect on the design of aerospace components. Load-bearing structures (e.g., booms, spars, vertical and horizontal stabilizers) typically are expected to possess tensile strengths of 100 MPa, moduli of 3,000 to 5,000 MPa, heat distortion temperatures in excess of 500 K and compressive and shear strengths of from 50 to 100 MPa. In these structural roles, both unfilled resins and fiber-reinforced resins display strains-at-break of 2% to 5% and low ($< 10^{-5}$) thermal expansion coefficients. A wide range of polymers is used in secondary structural roles for which moderate strengths (20-50 MPa) and moduli (500 to 1000 MPa) are required and strains-at-break of up to 5% are encountered. The temperature extremes to which these structures are exposed is narrower, from 150 to 350 K, approaching the glass transition or heat distortion temperatures. Although for most purposes these polymer materials can be considered as insulators, environments where exposure to ionizing radiation is severe require more careful distinctions to be drawn. Tolerance to exposures exceeding 10^9 rads without significant degradation is generally required for spacecraft materials. These distinctions are particularly important in the choice of materials whose primary function is to perform as an insulator or other dielectric material.

Polymers for thermal control include fairly specialized functions such as thermal insulators, ablative materials, and surface coatings to maximize the reflection of sunlight and the radiative cooling of the satellite. Thermal stability, heat capacity, and thermal conductivity are of primary concern in these materials. Mechanical properties are of secondary concern since these materials are usually applied as coatings. Because these materials are frequently exposed to temperature extremes and directly to the space environment, the absence of volatile, low molecular weight components is particularly important.

Low volatility is also a stringent requirement for materials used as lubricants, adhesives, and sealants. Because these materials are usually applied as a liquid or must remain soft during use, care must be taken to eliminate all volatiles in assembly. Often this can be achieved by complete curing or by heating under vacuum to remove residual solvent or monomers. The mechanical and thermal properties will vary considerably between applications. A structural adhesive for bonding metal parts must possess properties comparable to those for the secondary structural components. The properties of sealants and elastomers are generally very sensitive to temperature, so knowledge of the temperature extremes is essential in selecting a materials.

It is apparent that organic polymers are used in a wide variety of spacecraft applications requiring specific properties. However, all these materials are dielectrics and represent a potential source of electrostatic charging and electromagnetic interference; therefore, the designer must be aware of their electrical properties as well. The causes and consequences the spacecraft charging phenomena are discussed in detail in the proceedings of two conferences on the subject.[2,3] To aid in the selection or design of improved materials, we compiled a list of typical properties that materials used in these applications must possess. The properties and their reported values were compiled in Appendix A on the basis of function (e.g., strength retention over a given temperature range) and on the basis of distinctions drawn between materials that were successfully used and those that were rejected for a given application. To aid in the evaluation of new materials, the standardized ASTM test methods relevant to each property are compiled in Appendix B.

Commercial Materials Used in Spacecraft

The process for selecting spacecraft materials is similar to the process used to select materials for terrestrial use. The space environment, however, necessitates certain additional qualities. For example, the temperature extremes to which spacecraft materials are exposed can range from +250 to -250°C (excluding launch conditions). The presence of volatile components is extremely undesirable because they will "pollute" the spacecraft and its environs. Exposure to high energy and solar radiation is considerably higher than is commonly encountered in terrestrial applications. Furthermore, the materials must be durable because replacement of components is impossible if not prohibitively expensive and useful lives of up to ten years are expected.

In Appendices C and D we summarized the ratings of commercial plastics for use in structural applications and as dielectrics, respectively. As the ratings indicate, selections can seldom be based on absolute evaluations but frequently involve a tradeoff of relative strengths and weaknesses.

To provide a more specific standard against which contemporary and experimental materials may be judged, we have, in Appendix E, compiled a list of materials that are used in spacecraft or are candidates for use and evaluated their properties based on the materials requirements listed in Appendix A. Whenever possible, literature references are supplied; our estimates are unreferenced.

One interesting way to use these data is to compare the properties of materials already in use with those of polymers with unusual electrical properties. Note that very few of the properties relevant to space-based use have been measured for these conducting and semiconducting polymers. Experimental assessment of the use of these materials in reducing spacecraft charging will require that these properties be measured with some precision.

Our review of polymer use has indicated a wide range of applications and properties, yet a relatively limited number of materials are actually used. Generic description of the requirements of specific applications have been formulated. A range of contemporary materials and experimental polymers with unusual electrical conducting properties were evaluated according to those criteria. Few appear to be suitable for immediate off-the-shelf use, and more experimental data are needed to better evaluate the majority of semiconducting and conducting polymers for space-based applications.

References

1. J. Rittenhouse, J. Singletary, Space Materials Handbook, NASA-SP-3025, 1968 Third Edition, AFML-TR-68-205, AD 692353 (1966).

2. Proceedings of the Spacecraft Charging Technology Conference, C. P. Pike and R. R. Lovell, Eds., AFGL-TR-77-0051 (1977).

3. Spacecraft Charging Technology - 1978, R. C. Finke and C. P. Pike, Eds., AFGL-TR-79-0082 (1979).

4. Polymers in the Spacecraft Environment

Many spacecraft applications of polymers require their direct exposure to the space environment. The primary constituent of that environment responsible for spacecraft charging is high energy electrons although the effects of ultraviolet radiation, cosmic rays, and particles must be acknowledged. The consequences of this exposure and the control of charging were studied in several experimental missions (SCATHA, DSCS-III, ISEE-I, ATS-5, and ATS-6). In this section we summarize the interactions between the various types of high energy radiation and polymeric materials and the results of these interactions.

Effect of Radiation on Solid Polymers

Since the early work on the electron-induced vulcanization of rubber by Newton[1] in 1929, numerous and extensive studies have been conducted on the effects of radiation on polymers. We reviewed a large amount of the literature to correlate the available data, and in the following sections we summarize the radiation effects on polymers, then discuss the specific data available on the radiation-induced conductivity of various polymers.

The radiation-induced changes in the physicochemical properties of polymers are influenced by the type of radiation, the chemical structure of the polymer, the polymer's physical state during irradiation, and, the irradiation conditions (such as temperature, atmosphere, and radiation intensity).

Type of Radiation

Different types of radiation have been used to study the effect of radiation in polymers. These included x-rays and γ-rays, accelerated high-energy electrons, accelerated ions, and mixed radiation (γ-rays and neutrons) from nuclear reactors. Results of tests conducted under identical conditions led to the conclusion that, regardless of the type of radiation used, "equal amounts of energy absorbed produced equal changes in the properties of a polymer."[2] This is not entirely true, however, because it is not always possible to expose a single material to different types of radiation under identical conditions. For example, there is no difference in the reaction mechanism of high energy electrons with solid polymers and γ-rays with solid polymers. Both types of radiation lose most of the energy through ionization and excitation and cause very few displacement of atoms by recoils. However, the electron is completely stopped in the surface (~ 1 mm) of a relatively thick sample, whereas the photon will penetrate deeper into the material. Moreover, if oxygen is present, the electron effect, being concentrated on the surface of the sample, will be greatly enhanced. Because external conditions have such a pronounced influence on the nature of radiation effects, the assertion that equal amounts of absorbed energy produce equal changes applies only for changes produced by a single radiation type, except under very carefully controlled laboratory conditions.

Chemical Structure

The main effect of high energy radiation on organic polymers is to produce ionization and excitation. Subsequent rupture of chemical bonds yields polymer chain fragments, which may hold unpaired electrons from the broken bonds. The free radicals thus produced may react to change the chemical structure of the polymer and alter its physical properties. Depending on its chemical structure, the polymer may undergo crosslinking or scission.

Crosslinking increases the molecular weight of the polymer, decreases the solubility, and increases the softening temperature. As the radiation does increase , branched chains form and, finally, a tridimensional network results. As the density of the crosslinks increases, a soft, amorphous polymer will become rubbery, then hard and glassy.

In contrast, when the polymer suffers scission or degradation, its molecular weight steadily decreases with radiation dose, and the final product could be a low molecular weight liquid.

Both crosslinking and degradation may occur in the same polymer during irradiation, depending on its chemical structure. Polymers with tetrasubstituted or halogenated carbon atoms in the main chain, such as polyisobutylene and polyvinyl chloride, undergo scission; those containing at least one hydrogen on a carbon adjacent to a methylene group, such as polyethylene, tend to crosslink. Polymers containing aromatic substituents, such as polystyrene, resist radiolysis because of the deactivation produced by the aromatic group. Table 4-1 lists some common polymers that predominantly crosslink or degrade under irradiation.

Most polymers change colors during irradiation. These changes depend on the structure of the polymer, the irradiation temeprature, and the type of radiation. The color is also affected by trace amounts of additives pressent in the polymer and varies for irradiations performed in vacuum and in air.[3] The discoloration is attributted to the formation of conjugated double bonds in some polymers and also to trapped free radicals, electrons, and ions.

Gas evolution has been observed during irradiation of all polymers. The composition of the gas evolved varies with the nature of the polymer, the type of radiation, the dose rate, and the irradiation conditions. Hydrogen makes up 85% to 95% of the gas evolved from irradiation of polyethylene;[4] the other constituents are low molecular weight hydrocarbons. When bulky pieces of certain polymers are irradiated, the evolved gases often remain trapped in the polymer and produce an internal pressure. This internal pressure causes stresses and strain that lower the impact strength of the material. Sometimes the internal strain produces cracks in the sample and causes its final disintegration.

The electrical conductivity of polymers increases significantly under irradiation. The electrons and ions produced during irradiation can move in an electric field, and the current flowing is a measure of the radiation-induced current. This effect was first observed by Farmer

Table 4-1

SOME COMMON CROSSLINKING AND DEGRADING POLYMERS

Degrading Polymers	Crosslinking Polymers
Polyethylene $+CH_2-CH_2+$	Polyisobutylene CH_3 $+CH_2-C+$ CH_3
Polypropylene $+CH_2-CH+$ CH_3	Polymethacrylic esters CH_3 $+CH_2-C+$ $COOR$
Polyacrylic esters $+CH_2-CH+$ $COOR$	Polymethacrylamide CH_3 $+CH_2-C+$ $CONH_2$
Polyacrylamide $+CH_2-CH+$ $CONH_2$	
Polyvinyl chloride $+CH_2-CH+$ Cl	Polytetrafluoroethylene (Teflon) $+CF_2-CF_2+$
Polystyrene $+CH_2-CH+$ C_6H_5	Poly(vinylidene chloride) Cl $+CH_2-C+$ Cl
Polyethyleneterephthalate $+\overset{O}{\overset{\parallel}{C}}-\!\!\left\langle\!\!\bigcirc\!\!\right\rangle\!\!-\overset{O}{\overset{\parallel}{C}}-CH_2CH_2O+$	Poly(α-methylstyrene) CH_3 $+CH_2-C+$ C_6H_5
Nylon $+NH(CH_2)_6NH\overset{}{\underset{\parallel}{C}}+CH_2)_4\overset{}{\underset{\parallel}{C}}+$ OO	Polytrifluorochloroethylene Cl F $+C\!\!-\!\!C+$ FF

in polystyrene,[5] and thereafter it was studied in many other polymers.
It is assumed that conduction electrons become trapped in localized
traps, from which they can be released by thermal energy. The release
of the electrons can be measured by recording the current flowing
through the sample. The dependence of induced conductivity and the
subsequent decay on temperature and dose rate have been
investigated.[6,7] The photo-conductivity σ varies with dose I according
to the following relationship:

$$\sigma = RI^{\Delta}$$

where Δ has a value between 0.5 and 1 and that is characteristic of a
given substance and $\tilde{R}$ is a proportionality constant. A general theory
of induced conductivity in solid insulators was developed by Fowler.[7]
Materials with Δ = 1 show induced currents that are independent of the
temperature and that have short decay times; materials with Δ = 0.5 show
induced currents that increase rapidly with temperature and that have
much longer decay times. The theory indicates that the first class of
materials has a uniform distribution of traps and that the second class
has an exponential trap distribution.

Spark discharges are produced when polymer blocks are irradiated
with a uniform beam of energetic electrons. The electrons are stopped
in the block and a negative space charge builds up in the polymer. If a
grounded-point electrode is pressed against the sample, a spark
discharge occurs. For higher radiation doses, such discharges occur
spontaneously and are often initiated by local defects in the sample.
Under proper conditions, a characteristic discharge pattern builds up in
the polymer. This phenomenon has been studied by Gross[8] for
polymethylmethacrylate. Similar patterns have been observed in other
polymers on irradiation at low temperature.

Other Variables

Reactions involving free radicals are affected by anything that
changes the mobility of the species, such as temperature. The rate of

production of free radicals is independent of the mobility and dependent on the intensity of the radiation. The radical recombination reaction competes with other reactions such as crosslinking or scission. At a given radical production rate, the crosslinking rate is decreased by lowering the temperature. Both the rate of crosslinking in polyethylene and the rate of scission in polyisobutylene are strongly dependent on the temperature.[9,10] However, scission and crosslinking do not increase at the same rate with temperature.

If the irradiated polymer is partly crystalline, the free radicals formed within the crystalline regions remain firmly trapped because the mobility of the polymeric segments is low in these regions, particularly at low temperatures. The free radicals formed in the amorphous regions of the polymer can be readily used in chemical reactions, such as crosslinking or grafting, whereas those trapped in the crystalline areas can react only at elevated temperatures when most of the crystallites have melted.

The presence of oxygen during irradiation has a pronounced effect on the radiolysis of polymers. Materials that normally crosslink under irradiation can be degraded in the presence of oxygen. The effect depends on the physical state of the polymer. For example, thin films or powders of the polymer are greatly affected because the oxidation process is diffusion controlled. The radiation-induced degradation is also enhanced by the presence oxygen.

Commercial polymers may contain impurities and additives such as stabilizers, plasticizers, and antioxidants that can appreciably alter the radiation-induced changes. Active additives, such as those containing aromatic rings, act as energy sinks when incorporated into a polymer and improve the radiation resistance of the polymer. For example, the radiation-induced scission of polymethylmethacrylate is significantly reduced by incorporation of aromatic compounds, such as hydroxyquinoline and naphthalene, in the polymer.[11] The degree of protection is a function of the concentration of the additive, and this protection is attained without linking the additive to the polymer.

The inclusion of inert fillers does not alter the average dose received by the polymer binder, but aids in the retention of strength as the polymer binder degrades. Organic fillers, such as cellulose fibers, may deteriorate faster than the main polymer and accelerate its radiation damage. Similarly, many plasticizers that are sensitive to radiation enhance radiation damage of the base polymer.

Stabilization can also be built into the base polymer by copolymerizing an aromatic monomer with the major monomer constituent.

Conclusions

The chemical structure controls the behavior of polymers under irradiation. Crosslinking or scission is the major effect of radiation.

Crosslinking generally improves the physical properties of the polymer. It increases the tensile strength, Young's modulus, and softening temperature, and decreases solubility, elongation, elasticity, and viscous flow. It may cause embrittlement; it reduces crystallinity and therefore increases the light transmission of the polymer.

Degradation or scission has the opposite effect. It usually decreases the tensile strength, Young's modulus, hardness, and softening temperature and increases elongation and solubility.

Selective additives can minimize radiation damage. Nonpenetrating radiation can be used to harden the suface of a polymer, leaving the base soft and pliable. Another polymer can be grafted to the surface of the original polymer to modify its properties. Surface treatment with radiation may make a polymer more receptive to different types of dye.

Permanent changes in the electrical resistance of polymers can be produced by radiation damage. In addition, radiation yields ionic products, which reduce the electrical resistance by several orders of magnitude. During irradiation, electrons are excited to a free state, the conduction band, and produce photoconduction. The electrons may become loosely bound at trapping sites within the polymer and will affect the resistivity of the polymer for different periods of time after irradiation. These loosely bound, trapped electrons produce optical coloration similar to the color centers in ionic crystals.

Color changes are also attributed to the free radicals produced by the radiation-induced rupture of polymer molecules. In addition, color is produced in the polymer by the conjugated double bond systems formed during irradiation. These conjugated systems have electron excitation in the visible spectral range that causes discoloration and reduction of light transmission.

Radiation-Induced Conductivity in Polymers

As indicated in the preceding section, the electrical conductivity of polymer increases significantly on exposure to high energy irradiation. The study of this effect is of great importance, particularly in relation to polymers used as electrical insulators or in spacecraft where they are exposed to high energy irradiation. In the context of this report, however, radiation-induced conductivity is considered as a separate and distinct topic apart from polymer conductivity in general. It can be highly localized and transient and therefore is unlikely to provide relief from spacecraft charging phenomena.

The first systematic study on the radiation-induced conductivity in plastics was conducted by Fowler and Farmer.[12] Since then the radiation-induced conductivity in various polymers has been evaluated, and some of the data are presented below.

Polyethylene (PE)

The radiation chemistry of PE has been extensively studied.[13,14] Although its chemical formula indicated that PE is simply a long-chain molecule consisting of methylene groups, the structure of PE is rather complex. Conventional low density PE, prepared by free radical polymerization under high pressure, consists of long-chain molecules containing a few long branches and many short-chain branches. The PE prepared by the low pressure polymerization process is a linear unbranched polymer of high crystallinity (85-95%) and density. The low density PE is about 50% crystalline. Because of these differences in structure and crystallinity, the radiation effects for the two types of PE are not the same.

The presence of trapped electrons in γ-irradiated PE in the dark at 77 K was demonstrated by Keyser et al.[15,16] by electron spin resonance (ESR) studies. The concentration of trapped electrons in PE as a function of radiation dose was investigated by Keyser[7] for irradiation doses up to 6×10^{19} eV/g. The concentration increased to a maximum at 3×10^{19} eV/g and thereafter decreased with further increase in dose. At doses below 0.5×10^{19} eV/g, the trapped electron concentration increased linearly with dose, yielding a limiting value of $G(e_t^-) = 0.46$ trapped electron/100 eV at low doses for linear PE (Marlex 6050). The corresponding value for low density PE (Alathon 1414) was 0.12 trapped electron/100 eV. The fact that the linear PE showed a significantly higher $G(e_t^-)$ than the branched PE led Keyser et al.[15] and Keyser and Williams[16] to propose that the trapping site is associated with the crystalline regions of the PE. However, the nature of the electron trap (e.g., voids, holes, cavities in folded chains, or crystal defects) is still controversial.

The conductivity induced in PE by x-ray and γ-ray irradiation has been measured by Fowler and Farmer.[12,18] The equilibrium-induced current, i, in PE was dependent on the dose rate, R, according to the following relationship:

$$i \propto R^\Delta$$

The value of Δ was 0.8 ± 0.05 for PE over the range of R = 0.05 to 30 rad/min and between 20° and 80°C (x-rays of 220 KV were used). This value is in agreement with some of the Δ values obtained by other investigators as shown in Table 4-2. The activation energy for the radiation-induced current in PE was found to be 0.3 to 0.4 eV, and the corresponding activation energy for dark conduction or static conductivity was 1.5 eV. The induced current was almost independent of

Table 4-2: Radiation-Induced Conductivity in Polyethylene (PE)

Reference	Radiation Type	Experimental Conditions	Dose Rate Dependence, Δ	Induced Conductivity at Equilibrium $(\text{ohm cm})^{-1}$	Activation Energy E (eV)
19	X-rays	0.05–20 rad/min	0.8 ± 0.05 between 20°C and 80°C	10^{-16}	1.5 for static conductivity 1.4 for induced conductivity
20	X-rays	7 rad/min	0.8 ± 0.05 at 80°C	10^{-16}	
7	X-rays	10^8 rad/min at 20°C cable film cylinder condenser	0.81 0.81 ± 0.03 0.82 ± 0.03 0.79 ± 0.05	9.0×10^{-17} 1.2×10^{-16} 1.4×10^{-6} 10^{-16}	0.35
21	X-rays	10 rad/min at 20°C	0.75		
22	X-rays	10 rad/min at 20°C	0.75	10^{-16}	
23	X-rays	10 rad/min at 20°C	0.58	6.0×10^{-16}	
24	γ-rays	10^3–10^5 rad/hr at 190–300 K LPDE[b] film (1 mm thick) Temp °C: +9 +2 −20 −54 −75	0.61 0.61 0.61 0.83 0.83		0.34 0.34 0.34
		LDPE cable at room temp.	0.68		0.32
25[a]	γ-rays γ-rays	30–1220 rad/min room temp. 425 rad/min		10^{-14}	
26	γ-rays	5–500 rad/sec HPPE[c] 160 K 190 253 300 380 LPPE[d] 293	0.97 ± 0.03 0.96 0.83 0.84 0.78 0.74		

[a] Static conductivity range $5 - 20 \times 10^{-18}$ $(\text{ohm cm})^{-1}$.
[b] LPDE = Low density PE.
[c] HPPE = High pressure PE.
[d] LPPE = Low pressure PE.

the temperature below -40°C. Above this temperature the induced current
i increases with temperature, T, according to the relationship

$$i \propto exp(-E/kt)$$

In the low temperature range, the induced conductivity is due to charge
carriers, mainly electrons, activated by the irradiation; at the higher
temperature range, the thermally released trapped carriers contribute
significantly to the conductivity.

Polytetrafluoroethylene (PTFE) or Teflon

Teflon is one of the most inert materials, but it is extremely
sensitive to radiation. Its mechanical properties deteriorate rapidly
by irradiation in the presence of air; in the absence of oxygen, much
less radiation damage occurs.[26]

The radiation-induced conductivity in PTFE has been extensively
studied. Results of measurements made on the x-ray- and γ-ray-induced
conductivity in PTFE are shown in Table 4-3. PTFE compares with PE, in
the value of Δ increases at low temperature. Such variations in the
value of Δ are consistent with the model of exponential distribution of
traps in depth, postulated by Fowler.[7] The conductivity of Teflon was
found [25] to be greater than that of PE during irradiation and was found
to have a very long time-decay constant.

Polyfluoroethylene Propylene or Teflon FEP

Gross et al.[28] measured the conductivity induced in Teflon FEP by
x-rays at atmospheric pressure. The current induced by a radiation
intensity of 100 rads increases initially with time, reaches a maximum,
then decreases. A steady-state value is reached after 1 hr of
irradiation. Radiation-induced polarization or space-charge effects
could not be detected. After termination of irradiation, there is a
fast initial decrease of the induced conductivity, followed by a gradual
decay, similar to that described by Fowler.[7] This gradual decay is
influenced by the exposure preceding the measurement.

Table 4-3

RADIATION-INDUCED CONDUCTIVITY IN TEFLON POLYTETRAFLUOROETHYLENE (PTFE)

Reference	Radiation Type	Experimental Conditions	Dose Rate Dependence, Δ	Induced Conductivity (ohm cm)$^{-1}$	Comments
20	X-rays	7 rad/min at 20°C	0.62 ± 0.05	10^{-16}	Activation energy E = 0.5 eV
7	X-rays	8 rad/min at 20°C	0.63	8×10^{-17}	Variation in Δ with temperature
26	γ-rays	5–500 rad/sec, slightly oriented, 393 K	0.74		
		More oriented 293 K	0.87		
24	γ-rays	10^3–10^5 rad/hr at 190–300 K		10^{-14}	Activation energy E = 0.32 eV
		Temp °C			
		Film +9	0.83		
		−50	0.97		
		0.5 mm thick −61	0.97		
25	X-rays	30–1220 rad/min at room temp.		10^{-14}	No visible coloring
	γ-rays	425 rad/min at room temp			Static conductivity ranged from 5–20 x 10^{18}

The radiation-induced conductivity in films of Teflon FEP is significantly reduced after the film is exposed to massive doses (100 Mrad) of electron bombardment.[29] These high doses cause radiation hardening of Teflon. The radiation-hardened Teflon can store a considerably larger amount of negative charge in vacuum than the nonhardened material. This stored charge is released within a short time, when the irradiated sample is brought up to air pressure. The experiments were conducted in a high vacuum of 10^{-5} torr. This charge release occurs partly as internal discharges in the dielectric at pressures below 1 torr and partly as continuous charge release at higher pressures. The radiation hardening of Teflon ressembles similar effets observed in semiconductors.

The electron radiation-induced conductivity in Teflon FEP was also studied by Gross et al.[31] with partly penetrating electrons of a 40-keV beam and Teflon films 2.5×10^{-3} cm thick. The observed buildup and decay of the radiation-induced conductivity was explained by a "generalized box model" using a time-dependent conductivity $\sigma(t) = \sigma_o(1 - e)^{1/\tau}{}_s$ where σ_o is steady-state conductivity and τ_s is the time constant of the radiation-induced conductivity (RIC) buildup. The values of Δ and the RIC for Teflon FEP are shown in Table 4-4.

Polystyrene (PS)

PS is one of the most radiation-insensitive polymers; very large doses are required to produce any noticeable effect. A long time-constant of recovery for the x-ray-induced conductivity in PS was observed by Fowler and Farmer.[20] The dependence of the induced conductivity on temperature and dose rate during and after irradiation has been investigated.[7,26] Some of the values obtained are shown in Table 4-5.

The time-constant of decay varies slowly with temperature, showing the similarity between PS and PE and PTFE. At temperatures below $70^{\circ}C$ the current first falls rapidly, but after a few minutes it falls more slowly. The current remaining after a few minutes is proportional to 1/time, as for PE and PTFE. The magnitude of the induced current is

Table 4-4

RADIATION-INDUCED CONDUCTIVITY IN TEFLON POLYFLUOROETHYLENE PROPYLENE (FEP)

Reference	Radiation Type	Experimental Conditions	Dose Rate Dependence, Δ	Induced Conductivity $(\text{ohm cm})^{-1}$	Comments
28	X-rays	20-220 rad/sec	0.64	3×10^{-16}	Polyfluoroethylene-propylene foils
	γ-rays	1.25 MeV		1×10^{-14}	
31	e-Beam	40 keV Teflon FEP foils thickness 2.5×10^{-3} cm	0.79	5.9×10^{-15}	

Table 4-5

RADIATION-INDUCED CONDUCTIVITY IN POLYSTYRENE

Reference	Radiation Type	Experimental Conditions	Dose Rate Dependence Δ	Induced Conductivity $(ohm\ cm)^{-1}$	Comments
20	X-rays	7 rad/min at 80°C	0.60 ± 0.05	10^{-17}	E = 1.2 eV for static conductivity E = 0.44 eV for induced conductivity
6	X-rays	8 rad/min at 20°C	0.65	2×10^{-18}	
	X-rays	8 rad/min at 20°C	0.75	1×10^{-18}	
		U.S.A. sample			
26	γ-rays	5-500 rad/sec at 293 K	0.89		Atactic

smaller than for these two polymers.

Polyethylene Terephthalate (PET) or Mylar

In early studies on the radiation effects on PET, Charlesby[32] found that PET crosslinks at high radiation doses, whereas Todd[33] reported degradation; both irradiations were conducted in air. The predominant reaction is crosslinking, but because of PET's high concentration of phenyl groups, the radiation effects occur with a low yield.

The magnitudes of the static and x-ray-induced conductivities in PET have been determined.[7,34] The induced currents are smaller than those in the polymers described above (see Table 4-6). The equilibrium-induced current varies with dose rate following more nearly a linear than a square root law; when irradiation ceases, the current falls within a few seconds to a small fraction of the equilibrium value.

Similar results were obtained by Conrad and Marcus,[35] who measured the radiation-induced conductivity in a PET capacitor; the values are shown in Table 4-6.

Maeda et al.[36] studied the γ-ray-induced conductivity of PET under high electric field (1×10^6 to 1.6×10^8 V/m). At field strength below 10^8 V/m, the induced current reaches its initial steady state (primary component) soon after irradiation is started, and then it decreases with time. Above 10^8 V/m, the induced current decreases for a while and then increases to reach an equilibrium (secondary component). The radiation-induced conductivity of the primary component increases with electric field strength and then shows a saturation tendency. The current is proportional to dosage and can be analyzed into two components that behave differently with respect to time and voltage dependence. These are explained by space charge formation, by electron emission from the cathode at very high field strengths, by the influence of the field on initial recombination or by trapping.

Polymethylmethacrylate (PMMA)

The x-ray-induced conductivity in plasticized and unplasticized PMMA was studied by Fowler and Farmer.[7,20] The time constant of

Table 4-6

RADIATION-INDUCED CONDUCTIVITY IN POLYETHYLENE TEREPHTHALATE (PET)

Reference	Radiation Type	Experimental Conditions	Dose Rate Dependence, Δ	Induced Conductivity (ohm cm^{-1})	Comments
7	X-rays	8 rad/min at 20°C	0.83	6×10^{-20}	Variation in Δ with temperature. Basically suitable as a high insulation material.
34	X-rays	7 rad/min at 20°C R = 6–64 rad/min Temp = 20–100°C	0.83 ± 0.05 0.83 ± 0.05		
35	γ-rays	Source Flux (rad/sec) Fowler 1.3×10^{-1} Co-60 1.0×10^{2} Co-60 3.5×10^{3} DORF 6.5×10^{4} DORF 1.5×10^{5} DORF 3.3×10^{5} DORF 7.5×10^{5} DORF 3.8×10^{6}	0.83	6×10^{-20} 1.11×10^{-16} 5.65×10^{-15} 1.53×10^{-13} 3.69×10^{-13} 9.67×10^{-13} 1.88×10^{-13} 1.10×10^{-11}	DORF = Diamond Ordnance Radiation Facility (pulsed nuclear reactor).
36	γ-rays	Thickness (μm) / Electric field (V/m) 50 / 1×10^{6} 50 / 1×10^{7} 50 / 1×10^{8} 6 / 1.2×10^{8} 9 / 1.2×10^{8} 12 / 1.2×10^{8} 25 / 1.2×10^{8} 50 / 1.2×10^{8}	1.0 0.94 0.88 1.0 0.93 0.92 0.87 0.87	 1.92×10^{-16} 2.50×10^{-16} 3.33×10^{-16} 3.33×10^{-16} 3.67×10^{-16}	The Δ value is nearly the same for the 6-μm and 50-μm thick samples below 5×10^{7} V/m. Above this the Δ value increases, becoming close to unity for the 6-μm thick film, whereas for the 50-μm thick film it decreases monotonically with electric field. From 1.8×10^{8} V/m to 2×10^{8} V/m, Δ in the 6-μm thick sample was unity.

recovery after irradiation is very long (several hours) for the
unplasticized material, whereas for the plasticized PMMA it is much
shorter. The Δ constant for the unplasticized PMMA is 0.55, and for the
plasticized PMMA it is nearly 1 (see Table 4-7). Thus, the
unplasticized PMMA is similar in this respect to PE, PTFE, and PS.
Vaisberg et al.[26] studied the γ-ray-induced conductivity of plasticized
PMMA. The results were similar to those obtained by Fowler and Farmer.

Other Polymers

Examples of other polymers that have been studied for radiation-
induced conductivity are shown in Table 4-8.

Mixed copolymers, obtained by the polycondensation of hexamethylenediammonium adipate (15%), hexamethylenediammonium sebacate (40%), and
caprolactam (35%), were studied by Hedvig.[37] The conductivities induced
by x-ray and γ-ray irradiation were as high as 10^{-12} (ohm cm)$^{-1}$ with
dose rates of 0.1-20 rad/sec; no temperature dependence for the induced
conductivity could be detected between 10° and 60°C. The dark
conductivities followed the usual exponential rule with activation
energies of 1 and 2 eV. Preirradiation with γ-rays resulted in a
decrease in the induced conductivity, without appreciably affecting
the Δ value. Additives, such as benzophenone and hydroquinone,
influence the preirradiation effect by possibly controlling the radical
concentration. Hydroquinone, a strong radical acceptor, practically
eliminates the preirradiation effect, giving a rather radiation-
resistant material. A phenomenological model is presented to explain
the results.

The γ-rays induced conductivity in polycarbonate as a function of
the radiation dose rate was studied by Vaisberg.[26] The Δ value obtained
show the radiation-induced conductivity of PC is similar to PS.

The radiation-induced conductivity of polyimidazopyrrolone or
"pyrrone" polymers was studied by Reucroft et al.[38] (see Table 4-8).
The pyrrone polymers were prepared by the reaction of 3,3'-
diaminobenzidine (DAB) with either 3,3',4,4'-benzophenone

Table 4-7

RADIATION-INDUCED CONDUCTIVITY IN POLY(METHYL METHACRYLATE) (PMMA)

Reference	Radiation Type	Experimental Conditions	Dose Rate Dependence, Δ	Induced Conductivity $(ohm\ cm)^{-1}$	Comments
20	X-rays	7 rad/min at 20°C to 80°C	0.55 ± 0.05	10^{-17}	Unplasticized E = 1.6 eV[a] E = 0.6 eV[b]
7	X-rays	8 rad/min at 20°C	0.55	2×10^{-18}	Unplasticized Variation in Δ with temperature
	X-rays	8 rad/min at 20°C	1.00	3×10^{-18}	Plasticized
26	γ-rays	5–500 rad/sec at 293 K	0.88		Plasticized

[a] Static conductivity.
[b] Induced conductivity.

Table 4-8: Radiation-Induced Conductivity in Other Polymers

Reference	Polymer	Radiation Type	Experimental Conditions	Dose Rate Dependence, A	Induced Conductivity (ohm cm)$^{-1}$	Comments
36	Polyamide Copolymers	X-rays and γ-rays	200 kV X-ray unit and some with 50 Ci Co^{60} source β rays filtered out. Dose rates = 0.1–20 r/s	0.6	10^{-12}	No change in σ by changing temperature up to 60°C. Preirradiation with γ-rays resulted in a decrease in σ but the Δ value did not change appreciably.
25	Polycarbonate	γ-rays	5–500 rad/sec at 293 K	0.67		
30	Pyrrone Polymers (Copolymer of BTDA or PMDA with DAB)	γ-rays 120 rad/min and e$^-$ from a 2-MeV source	BTDA-DAB PMDA-DAB 10- to 50-μm-thick films DAB Applied field strengths = 10^2–10^6 V/cm 4,000 - Ci γ-beam 150 C irradiation facility Dose rates = 10–1,000 rad/min	0.5 0.5	6.25 × 10^{-16}	Radiation-induced current for BTDA-DAB was ≅ 3–4 times smaller than that in PMDA-DAB. In general, irradiated samples showed a higher dark current. Observed increase in BTDA-DAB much less than in PMDA-DAB. RIC effects and permanent effects much less marked in BTDA-DAB.

BTDA = Sublimed 3,3',4,4'-benzophenone tetracarboxylic acid dianhydride.
DAB = 3,3'-diaminobenzidine.
PMDA = Pyromellitic dianhydride.
RIC = Radiation-induced conductivity.

tetracarboxylic acid dianhydride or with pyromellitic dianhydride.
These polymers have shown excellent thermal and radiolytic stability.
Their mechanical properties were practically unchanged by accumulated
radiation doses to 10^{10} rad. The γ-radiation-induced conductivities in
pyrrone samples at dose rates up to 900 rad/min were generally lower
than those reported for common insulating polymers. The radiation-
induced current for BTDA-DAB was about 3 to 4 times smaller than that in
PMDA-DAB. Permanent increases in dark conductivity were produced in the
polymers by accumulated doses of 2-MeV electrons ranging from 1 x 10^7
rad to 5 x 10^9 rad at temperatures up to 300°C. The dark conductivity
increases produced were not sufficient to inhibit the UV-visible
photoconductivity of the polymers. No significant effect on the
dielectric properties of the polymers was detected by such accumulated
doses. Both radiation effects, the induced conductivity and the
permanent dark conductivity increase, are more pronounced in PMDA-DAB
than in PBDA-DAB.

Conclusions

The general effects observed in insulating polymers by exposure to
high energy irradiation are induced conductivity and permanent
conductivity changes caused by chemical changes in the polymer. Because
the radiation-induced conductivity generally is not sufficiently large
and because it usually does not extend into the adjacent nonirradiated
volume of polymer or cannot depend on radiation-induced conductivity to
leak off charge and prevent breakdown pulses.

The induced conductivity is related to the radiation dose rate R
according to the following relationship: $\sigma \propto R^{\Delta}$. The Δ exponent has
limits of $0.5 < \Delta < 1$ according to the Rose-Fowler model and remains
constant over a wide range of radiation doses. We find conflicting
results in the data available on Δ values determined for a number of
polymers by various investigators. The variations in the values
reported have been attributed to impurities in the polymers, differenes
in the physical state of the polymer, presence of trapped gases, and the
experimental conditions used. It is difficult to draw conclusions about

Table 4-9

DETERMINATION OF CARRIER MOBILITY IN INSULATING POLYMERS

Reference	Polymer	Temperature	Carrier Mobility μ (cm^2/V sec)
38	Linear polyethylene (Rigidex)	80°C 95°C	4.5×10^{-10} 9.0×10^{-10} Hole mobility
39	Polyethylene	Using surface charge decay technique	$10^{-7} \sim 10^{-10}$
38	Polyethylene	Room temp. pulsed e-beam	4.5×10^{-10} Hole mobility
40	Polyethylene	Pulse voltage technique	2.5×10^{-14}
38	Atatic polystyrene (Styrafoil)	20°C 80°C	1×10^{-6} 5×10^{-5} Hole mobility
38	Polystyrene		1×10^{-6} Hole mobility
41	Polystyrene	Room temp.	1×10^{-4} Electron mobility; 7×10^{-5} Hole mobility
38	Polyethylene-terephthalate (Melinex)	20°C 80°C	1.5×10^{-6} Electron mobility 2×10^{-5}
38	Polyethylene-terephthalate (PET)		1.5×10^{-6} Electron mobility
41	Polyethylene-terephthalate (PET)	Room temp.	2×10^{-5} Electron mobility, 1×10^{-4} Hole mobility
41	Polyethylene-naphthalate (PEN)	Room temp.	2×10^{-4} Electron mobility, 6×10^{-5} Hole mobility
41	Fluoroethylene propylene (FEP)	Room temp.	5×10^{-5} Electron mobility, 5×10^{-4} Hole mobility

the actual mechanism of the radiation-induced conductivity from the available data.

Carrier mobility in various insulating polymers has been studied to identify the sign of the carriers predominantly responsible for the radiation-induced conductivity in the materials and to obtain data useful in the interpretation of the radiation reaction and the factors influencing it. The results obtained by various investigators are presented in Table 4-9. The discrepancy in the values of carrier mobility obtained by the various workers is attributed to the measuring methods used.

References

1.   E. B. Newton, U. S. Patent 1,906,402 (1929).

2.   A. Chapiro, Radiation Chemistry of Polymeric Systems (Interscience Publishers, NY., 1962).

3.   A. Chapiro, J. Chim. Phys. 53, 895 (1956).

4.   M. Dole, C. D. Keeling, and D. G. Rose, J. Am. Chem. Soc. 76(17), 4304 (1954).

5.   F. T. Farmer, Nature 150, 521 (1942).

6.   R. A Meyer, F. L. Bouquet, and R. S. Alger, J. Appl. Phys. 27, 1012 (1956).

7.   J. F. Fowler, Proc. Roy. Soc. (London) A236, 464 (1956).

8.   B. Gross, J. Polymer Sci. 27, 135 (1958).

9.   A. Charlesby, Chem. & Ind. (London) 232, 1956(8) (1957).

10.   T. F. Williams and M. Dole, J. Am. Chem. Soc. 91(12),2919 (1959).

11.   P. Alexander, and D. J. Tomas, J. Polym. Sci. 22(101), 343 (1956).

12.   J. F. Fowler and F. T. Farmer, Nature 171, 1020 (1953).

13.   M. Dole, Rep. Symp. 9th Chem. Phys. Radiat. Dosimet. Army Chemical Center (1950).

14.   A. Charlesby, Proc. Roy. Soc. London A215, 187 (1952).

15. R. M. Keyser, K. Tsuji, and F. Williams, Macromolecules 1, 289 (1968).

16. R. M. Keyser, and F. Williams, J. Phys. Chem. 73, 1623 (1969).

17. R. M. Keyser, "An Electron Spin Resonance Study of Trapped Electrons in Gamma-Irradiated Hydrocarbon Polymers," Ph.D. Disertation, Univ. of Tennessee, Knoxville, Tennessee (1967).

18. J. F. Fowler and F. T. Farmer, Nature 174, 136 (1954).

19. J. F. Fowler and F. T. Farmer, Nature 173, 317 (1954).

20. J. F. Fowler and F. T. Farmer, Nature 175, 516 (1951).

21. W. L. Laurence and S. Mayburg, J. Appl. Phys. 23, 1006 (1952).

22. J. H. Coleman, Nat. Acad. Sci., Wash. 23rd Annual Meeting Conference on Electrical Insulation.

23. D. K. Keel, S. Koganoff, C. H. Mayhew, and H. G. Nordlin, USAEC Report No. NYD-4518 (1953).

24. K. Yahagi and A. Danno, J. Appl. Phys. 34 (1), 804 (1963).

25. R. A. Meyer, F. L. Bouquet, and R. S. Alger, J. Appl. Phys. 27, 1012 (1956).

26. S. E. Vaisberg, V. P. Sichkar, and V. L. Karpov, Translated in Polym. Sci. USSR, A13, 2812 (1972)

27. L. A. Wall and R. E. Florin, J. Appl. Polym. Sci. 28, 653 (1958).

28. B. Gross, R. M. Faria, and G. F. Leal Ferreira, J. Appl. Phys. 52(2), 571 (1981).

29. B. Gross, G. M. Sessler, and J. E. West, Appl. Phys. Lett. 24, 351 (1974).

30. J. P. Maita, Phys. Rev. 93, 693 (1954).

31. B. Gross, J. E. Wests, H. Von Seggern, D. A. Berkeley, J. Appl. Phys. 51(9), 4875 (1980).

32. A. Charlesby, Nature 171, 167 (1953).

33. A. Todd, Nature 174, 613 (1954).

34. J. F. Fowler and F. T. Farmer, Nature 175, 590 (1955).

35. E. E. Conrad and S. M. Marcus, DTIC Technical Report. TR-1037, 4 May 1962.

36. H. Maeda, M. Jurashige, and T. Nakakita, J. Appl. Phys. $\underline{50}$(2), 758 (1978).

37. P. Hedvig, J. Polym. Sci. A, $\underline{2}$, 4097 (1964).

38. P. J. Reucroft, H. Scott, P. L. Kronick, and F. L. Serafin, J. Appl. Polym. Sci. $\underline{14}$, 1361 (1970).

39. E. H. Martin and J. Hirsh, J. Appl. Phys. $\underline{43}$, 1001 (1972).

40. D. K. Davies, J. Phys. $\underline{D5}$, 162 (1972).

41. T. Tanaka, J. Appl. Phys. $\underline{44}$, 2430 (1973).

42. K. Hayashi, K. Yoshino, and Y. Inuishi, Japanese J. Appl. Phys. $\underline{14}$(1), 39 (1975).

5. Survey of Semiconducting and Conducting Polymers

To identify specific materials with the electrical properties believed to be necessary for the reduction of spacecraft charging and the common molecular structures responsible for those properties, we have compiled an extensive list of resistivity data for both commercial and experimental polymers. Included are materials already used in spacecraft and those with resistivities on the order of 10^{12} ohm cm or less. These data, along with literature references and representative molecular structures, can be found in appendices E through G. Included in Appendix E are comments concerning the experimental conditions under which measurements were made, observations about the dependence of conductivity on temperature or radiation exposure, and miscellaneous remarks concerning the use of dopants, polymer crystallinity, and other material variations.

6. Commercial and Experimental Polymers for Use in Space

In this section we review physical property data on various polymer systems that may mitigate the problem of spacecraft charging. We adopt the following three requirements as being necessary for successful applications. First, the polymer should have a conductivity of about 10^{-12} (ohm cm)$^{-1}$. The optimum value may be different and may vary with specific applications, but will be near this intermediate value where radiation-induced charge can be drawn off before electric fields high enough to cause dielectric breakdown can accumulate.

Second, the material must be stable in a space environment. Temperature ranges from +250°C to -250°C may be encountered. Constant exposure to high energy electrons and cosmic radiation must be tolerated as well as exposure to hard vacuum. Some time element should be recognized as well. All payloads will have a finite lifetime depending on their orbit, power supply, and mission. We will aim for a ten-year useful lifetime, but this is probably an upper limit.

Third, the polymers must have the physical properties required for their primary function, e.g., structural members, thermal control, or packaging. These will vary considerably with different applications, but in discussing specific materials we will indicate what they are suitable for.

The commonly used industrial plastics are discussed first, followed by materials that have unusual electrical properties.

Polyimides (Kapton)

The polyimides are a general class of polymers containing the aromatic imide linkage. Almost all polyimide conductivity studies have been performed with Kapton, a commercial product widely used in the aerospace community. The conductivity is variable, depending on temperature, field strength, frequency and sample history[1,2] with values of 10^{-7} to 10^{-14} (ohm cm)$^{-1}$ reported. The exact mechanism of conductivity is uncertain, but ionic conduction at low fields is strongly indicated[2] by the temperature and field dependence of the current density. At high fields (100-500 kV cm^{-1}), thermally excited excitonic carriers may contribute to the conductivity. For spacecraft applications, the low end of the conductivity range should be used because water or other ionic species will be rapidly eliminated in this environment.

Kapton is a highly amorphous material with a significant potential for local structural disorder because it is prepared via an amorphous polyamic acid precursor. It does have good high temperature stability (T_g > 250°C) and mechanical properties that make it very useful as a high temperature coating or film. Overall, Kapton film meets all the thermal and mechanical requirements for dielectric, thermal control, and secondary structural element applications.

Various polyimide molding compounds and matrix resins for fiber-reinforced composites also exist, but they generally do not have the thermal stability and mechanical strength of Kapton films.

The polyimides have many desirable features and are close to the desired conductivity level of 10^{-12} (ohm cm)$^{-1}$. The polyimides could be modified by either creating more carriers or increasing the mobility of the carriers present.

Adding "molecular dopants," e.g., triphenylamine or trinitrofluorenone (TNF), that are readily ionized into localized states for hopping transport in polycarbonate would probably raise the conductivity by two or three orders of magnitude.[3] Molecular dopants form "conducting pathways" in a hopping analogy to carbon-filled systems. The use of molecular dopants will probably allow the realization of 10^{-12} (ohm cm)$^{-1}$ in almost all polymers.[4]

Unfortunately, molecular dopants are volatile and will plasticize the polymer. The consequent loss of mechanical properties and limited stability to thermal, ultraviolet, and high energy radiation precludes their use. The use of strong oxidizing dopants that enhance the conductivity of polyacetylene (AsF_5 or I_2) is unlikely to raise the electronic conductivity of polyimides significantly. The polyimides lack the high degree of structural order necessary for extensive intermolecular π bond overlap that the polyacetylenes have.

Very significant changes in the conductivity of Kapton, from 10^{-18} to 10^2 (ohm cm)$^{-1}$, occur when the material is pyrolyzed at 600° to 850°C.[5] Conduction in these pyrolyzed materials is electronic and thermally activated. In the pyrolysis process considerable quantities of nitrogen and oxygen gas are driven off, and the aromatic rings condense to form extended π molecular orbitals. The polymer density also increases as these extended two-dimensional networks stack more closely together than is possible for the linear polymer. Thus, the pyrolysis leads to both the formation of more carriers (unpaired electrons in the planar array) and an increase in their mobility through increased ordering and intermolecular (interplanar) π bond overlap. This is the first of many examples of how increased intermolecular order leads to better π electron delocalization and higher conductivity levels.

Although pyrolysis need not seriously degrade the strength or modulus of the polyimide, it may reduce the elongation at break to 1-2% and greatly reduce the flexibility of films or coatings. The molecular structure of the pyrolysis products is relatively uncertain and may be heterogeneous, particularly for lightly pyrolyzed materials. However, the good correlation between pyrolysis temperature and conductivity suggest that materials with very well-defined conductivities could be obtained.

Fluorocarbons (Teflon)

The fluorocarbons in general are highly insulating and resistant to thermal degradation, but they are are not suitable for structural applications. Their saturated aliphatic backbone is probably the least likely candidate for generating or stabilizing charge carriers.

However, they can be filled with conducting metal or carbon particles to produce low-friction conductivity surfaces. They remain useful in dielectric applications, but are unlikely to be useful in structural applications.

Polyesters (PET, PBT, Mylar)

As is the case for polyimides conductivity in the polyesters and the polyamides is almost always ionic and due to the presence of water.[6] Exposure to high energy radiation and large electric field strengths will produce excitonic carriers but at concentrations below that useful for mitigating the effects of spacecraft charging.

The electronic structures of commercially important polyesters tend to be very highly localized. Copolymers containing large delocalized ring structures could be prepared, but the loss of stereoregularity would drastically reduce the crystallinity and hence lower the materials' thermal and mechanical properties.

Several attempts have been made to use the polyester backbone as a carrier for introducing radical ion salts such as TTF and TCNQ[7] into a polymer matrix.[8,9] Similar polymers using polyurethane,[10] polysulfonate,[9] and polyethyleneimines[11] have been prepared. In general, the polymers produced have been of low molecular weight, insoluble, and amorphous. Little or no evidence for the formation of molecular complexes responsible for conduction in TTF:TCNQ salts has been observed in these polymers.

Many charge transfer complexes between polymer and an electron acceptor have been prepared or attempted.[12] Electron acceptors[8] include TCNQ, I_2, and TNF. Several moderately successful elastomeric materials based on quartenary polyurethanes[13,14] and quartenary-substituted vinyl polymers[15] are known. Conductivities of the former range from 10^{-16} to 10^{-17} (ohm cm)$^{-1}$ depending on temperature and composition. The elastomers have useful properties comparable to those of conventional polyurethane elastomers. The most interesting feature of these materials is that they develop conduction anisotropy on uniaxial elongation. They are not suitable for use as structural materials or in high temperature environments, but they could be very useful for applications requiring a moderately conductive vibration-damping material or potting compound. Conduction follows a simple thermally activated temperature dependence and the inverse-square-root dependence, of pressure indicates that conduction is not ionic.

Note that of the polymers reported in references 8 through 10, those with a more regular backbone structure and few pendant groups, two features that promote intermolecular order, have higher conductivities.

Poly(alkanes) (Polyethylene, Polypropylene)

The chemical composition of these materials is similar to that of wax, as are their electrical properties. They tend to be highly crystalline and inert to ionizing radiation. It is unlikely that any modification, short of adding conducting filler particles, will significantly increase their electronic conductivity. Ionic conductivity is observed in metal ion salts of ionomers. The toughness and elevated glass transition temperature make them very useful in films and coatings.

Vinyl Polymers and Acrylates

These are materials prepared from the free radial or anionic polymerization of vinyl or acrylic compounds. The most common examples are polystyrene and polymethylmethacrylate, respectively. The class includes many polymers, but we are considering them together for the following reasons. First, they melt far below 250°C, often becoming rubbery below 100°C, and are not widely used in spacecraft. Second, there is nothing about their backbone structure that, in our view or in the literature, is likely to lead to appreciable conductivity. Any useful conducting properties for those materials must be derived from pendant groups attached to the chain backbone. In general, however, large bulky electron systems will make polymerization and processing difficult. There are a few materials with noteworthy properties.

Polyvinylcarbazole (PVK)

Polyvinylcarbazole (PVK)[5] is best known for its photoconducting properties. It also has a fairly substantial dark conductivity ranging from 10^{-16} to 10^{-5}(ohm cm)$^{-1}$, depending on dopant and temperature.[16] Various dopants, including iodine (which will desorb under vacuum), TCNQ, and TNF, have been used.[12] PVK's mechanical and thermal properties make it a very good candidate for spacecraft applications. It is processable, very strong, and has a high T_g (150-200°C) depending on the degree of stereoregularity.

Mechanistically, conduction in PVK is similar to that observed in molecularly doped polycarbonate or TTF-substituted polyesters. The conductivity shows a simple inverse temperature dependence and increases gradually with dopant concentration to a maximum at a 1:1 mole ratio of dopant to vinyl carbazole.

The molecular structure of PVK clearly illustrates the two features that we believe are important in polymer conductivity. First, the carbazole ring structure possesses the minimum molecular structure required for stabilizing a charged species; the lone pair electrons on the nitrogen atom can be ionized relatively easily. Second, the structural order necessary for diffusing charge over several molecular residues is present. Steric repulsion requires that the large planar carbazole units be arranged normal to the backbone axis, and there is enough ring overlap for transport to occur with a relatively low activation energy.

There are ways to modify the conductivity of PVK. It can be readily copolymerized with a variety of vinyl monomers and doped with organic or inorganic materials. Moreover, the mechanical properties of PVK are as good as those of unpyrolyzed Kapton, and the development of a more highly crystalline or crosslinked modification of PVK would probably raise the glass transition temperature by 50° to 100°C. PVK films can be readily oriented by stretching, which enhances both the local structural order and mechanical properties such as strength and modulus.

Polyvinyl Pyridines

The properties of the vinyl pyridines are similar to those of PVK, a dopant (I_2) or molecular complexing agent (TCNQ) must be a strong enough oxidizing agent to remove an electron from the pyridine ring to bring the conductivity up to 10^{-3} (ohm cm)$^{-1}$ from the undoped value[15] of 10^{-15} (ohm cm)$^{-1}$. Conductivity in these systems follows the same thermal activation observed in other molecularly doped systems. Similarly the mole fraction of dopant is increased to approximately 1:1.

Application to spacecraft-charging problems would require the use of a nonvolatile dopant, possibly some sort of a copolymer, and an increased glass transition temperature. Like polystyrene, the polyvinyl pyridines soften at about 100°C, making them unsuitable without further modification for application as structural elements in spacecraft. Either increasing the stereoregularity or crosslinking the polymer after it is formed could raise the maximum use temperature.

Phthalocyanine-Containing Polymers

This general structural type of material contains aromatic phthalocyanine ring structures in the polymer backbone. Frequently, they are complexed with a metal ion, e.g., Al, Mn, or Fe, and may be prepared in either an idealized flat planar structure[17,18] or a cofacial assembly like doughnuts threaded onto a shaft.[19] Conductivity in all these systems is due to thermally activated hopping between adjacent sites and varies from 10^{-12} to 10^1 (ohm cm)$^{-1}$, depending on composition and doping. The planar, crosslinked systems do not require any dopant in addition to the metal atom complexed in each phthalocyanine ring. Conductivity in these systems is approximately proportional to the square root of the applied pressure, as would be expected for this type of system.[20] These are generally infusible, insoluble, amorphous solids, which makes processing extremely difficult.

However, the conductivity of cofacial or bridge-stacked compounds depends strongly on doping. Iodine is the most widely studied dopant, although other organic electron acceptors have been used. Conduction rises gradually as the mole ratio of dopant to metal ion reaches 1:1[21] as is observed for almost all of the "molecularly doped" charge transfer systems. They are also slightly more tractable than the crosslinked phthalocyanines. They can be dissolved in sulfuric acid and sublimed into thin films.[22,23]

No mechanical properties have been reported for these materials. They do have good chemical and thermal stability and in the undoped or partially doped state could be candidates for spacecraft applications. Although current work is aimed at optimizing conductivity, recognition of applications requiring polymeric materials with intermediate conductivities, e.g., spacecraft charging or electromagnetic interference (EMI) shielding, could direct attention toward measuring and optimizing mechanical properties.

Similar polymers have been obtained from condensing organic nitriles[24] and pyrolyzing polyacrylonitrile.[25] Conductivities of the former, which are low molecular weight dark-colored powders, range from 10^{-12} to $\sim 10^{-5}$ (ohm cm)$^{-1}$, but again the polymers cannot be adequately processed to form useful devices. Pyrolyzed polyacrylonitrile is prepared from films or fibers, but is relatively brittle and intractable after pyrolysis at 435°C.

As for the polyimides, it is difficult to assess whether or not a partial pyrolysis to increase the conductivity to $\sim 10^{-12}$ (ohm cm)$^{-1}$ without excessively degrading mechanical properties could be done. However, the conductivity does increase significantly from 10^{-9} to 10^{1} (ohm-cm)$^{-1}$, and volatile molecular dopants or oxidizing agents are not required. The fused ring systems formed are reminiscent of those formed during the pyrolysis of polyimides, and it would not be surprising if the polymer density and crystallinity increased on pyrolysis. The temperature dependence of the conductivity can nearly be described as a thermally activated process. However, as is pointed out in reference 25, experimental data over a narrow temperature range fit a variable range hopping model ($\ln \sigma \sim T^{-1/4}$) as well as the simple nearest-neighbor, thermally activated, hopping model.

Poly Acene Quinones

These materials have properties very similar to those of the poly-phthalocyanines.[26] Observed conductivities range from 10^{-11} to 10^{-2} (ohm cm)$^{-1}$. They can only be prepared in the form of insoluble, infusible, black powders of uncertain compositions, so their use is extremely limited.

However, these materials have some unique properties. First, they have an extremely high, dielectric constant. Typical values are from 30 to 10,000,[27] which is believed to be due to the extremely high polarizability of electrons in the delocalized π electron molecular orbitals. Second, no dopant effect is observed. Evidently, there is sufficient intermolecular order among the delocalized molecular orbitals for a variable-range hopping mechanism to appear. Both DC and AC conductivity measurements show a clear $\ln\sigma \sim T^{-1/4}$ behavior.[28] This is different from the simple thermally activated ($\ln\sigma \sim T^{-1}$) process where charge transport occurs between adjacent, evenly spaced, single-molecule localized states. As the name "variable-range" implies, charge transport occurs between states that may be spatially distant but otherwise favorably located energetically.

Metal-Containing Coordination Polymers

A wide range of conductivity values have been reported for metal coordination polymers, the most widely studied of which is polyvinyl ferrocene.[29] In these materials the metal ion may be coordinated either in the backbone or in a pendant group. Thermally activated hopping of electrons between metal ions of mixed oxidation states, e.g., Fe^{+2} and Fe^{+3}, is the most likely conduction mechanism. For example, the unoxidized (Fe^{2+}) polymer has a conductivity of only 5×10^{-12} (ohm cm)$^{-1}$. Oxidizing 73% of the iron to Fe^{3+} raises that value to 2×10^{-6} (ohm cm)$^{-1}$.[30] In comparison, complexing with TCNQ raises the conductivity ferrocene polymers to 2.5×10^{-5} (ohm cm)$^{-1}$, and an unsaturated vinylene backbone polymer complexed with I_2 has a conductivity of 10^{-6} (ohm cm)$^{-1}$.[31]

The physical properties of these materials are largely uncharacterized. Most do melt and are soluble in organic solvents, but because their conductivities, 10^{-13} to 10^{-2} (ohm cm)$^{-1}$, are not in the metallic regime, they have not been further pursued. Some do possess the requisite conductivity, probable thermal stability, and processability for possible application to spacecraft-charging problems.

Semiconducting Polymers with Conjugated Backbones

This section includes polymers with conjugated backbones and conductivities in the semiconducting range. We put these materials together to emphasize how unsuccessful it has been to model the bulk conductivity of a polymer according to the structure of the molecular repeat unit alone. For all these materials it is possible to draw delocalized π molecular orbitals that imply complete electron delocalization, as is found in a benzene ring. Many of these compounds are very close analogs to polyacetylene, which is highly conducting when fully doped, possessing the same π electron conjugated backbone. However, none of the analogs are highly conducting, even when doped. If nothing else, this literature survey teaches one important lesson: a conjugated π orbital backbone is neither necessary nor sufficient for conductivities of aoubt 10^0 (ohm cm)$^{-1}$.

Generally speaking, these materials are very difficult to process so very little mechanical property data are available. The structures are generally highly aromatic, which lowers the solubility and increases the melting point. Consequently, the molecular weights are low because the growing polymer chain cannot be kept in solution long enough to achieve a high molecular weight. Some, in particular the polybenzimidazoles,[32] polybenzothiazoles,[33] azomethines[34] and Schiff base polymers,[35] are soluble only in strong acids and can be spun into fibers with excellent thermal-oxidative stability.

Unfortunately, their resistance to oxidation is achieved by tight binding of electrons. Consequently, it is difficult to envision facile charge transport. Unless relatively low energy anionic or cationic states can be formed, the number, mobility, and stability of carriers will be insufficient for high levels of conductivity. Typical values in these systems range from 10^{-16} to 10^{-12} (ohm cm)$^{-1}$.

Other heterocyclic aromatic polymers, including the emeralidines,[36] azines,[37] and azophenylenes[38] have relatively good thermal stability and reasonably high glass transition temperatures and semiconducting properties of about 10^{-4} (ionic), 10^{-10} to 10^{-5}, and 10^{-12} (ohm cm)$^{-1}$, respectively, but have either not been prepared in reasonably high molecular weights or are intractable for practical purposes.

A subset of these conjugated backbone polymers are the analogs of polyacetylene. In general, they are more likely to be soluble in organic solvents, and they soften well below 200°C. Their thermal stability is not quite as high as the previously mentioned polymers, and their conjugated backbones are readily oxidized (or doped) by iodine. For some, iodine doping raises the conductivity by up to six orders of magnitude, but the temperature dependence of the conductivity is still indicative of thermally activated hopping between adjacent localized sites.[39]

Very little work has been done to characterize the mechanical properties of these materials because they do not have high levels of conductivity, high molecular weight, or superior thermal properties.

The most extensive characterization work on these materials has been performed on the substituted diacetylenes.[40] These are soluble in organic solvents and can be cast into tough films with strengths and moduli comparable to those of polyethylene. However, the glass transition temperature is relatively low, $\sim$0°C, and these materials melt sharply at 167°C.

Because both single-crystal and amorphous specimens can be prepared, the effect of intermolecular order is readily apparent. Both morphologies readily absorb iodine. In single crystals the adsorption of 2.6 mol% iodine raises the conductivity by seven orders of magnitude to nearly 10^{-6} (ohm-cm)$^{-1}$, whereas in an amorphous film nearly 40 mol% is required for a two-order-of-magnitude increase to the same conductivity level. In this system, for the first time, the dependence of conductivity and activation energy on molecular weight has been demonstrated. At constant dopant levels the conductivity increases linearly with molecular weight, and the activation energy is proportional to the reciprocal molecular weight, yielding an infinite molecular weight value of 0.44 eV. These are significant results: they clearly demonstrate that bulk conductivity relies on interchain transport and that models based on the structure of single repeat units or individual chains are incomplete.

The polymers studied contain about 35 repeat units. In single-crystal samples, where chain axis are well aligned and intrachain delocalization is maximized, the maximum conductivity is reached at about 2.5 mol% iodine or approximately one dopant molecule per polymer molecule.

In cast films the structure is largely amorphous, and the high dopant level at which the maximum conductivity is obtained corresponds to about one dopant molecule per 2.5 repeat units. One further observation concerning the dependence of conductivity on dopant concentration can be drawn from this system.

At very low dopant concentrations the conductivity is relatively insensitive to dopant. The conductivity rises sharply at a certain dopant concentration and then levels out. This same stepwise (nearly) increase in conductivity is also observed for the polyphenylene vinylene[41] and each of the highly conducting polymers discussed in the next section. The only system where this rapid increase has not been observed is iodine-doped polyphenylacetylene, for which conductivity is known to be ionic.[42]

Obviously, there are numerous semiconducting compounds in this class. It is unlikely that highly conducting polymers will be prepared from this group, primarily because high conductivity requires a high mobility. Mobilities, in turn, require a very regularly spaced molecular structure containing an abundance of readily ionized units. All conceivable side groups examined have significantly lowered the properties of highly conducting polyacetylene and polypyrrole. Massive pendant groups in substituted polydiacetylenes, although necessary for polymerization into single crystals, significantly lower the overall conductivity even though excellent intrachain delocalization is observed. The best approach to developing processable, highly conducting, conjugated-backbone chain polymers might involve the coupling of fairly large, symmetric, charge-stabilizing repeat units with short, flexible, readily soluble units.

Numerous opportunities exist, however, for application to the spacecraft-charging problem. Some of the thermally stable aromatic heterocycles are already used or being contemplated for use in aerospace vehicles. The problems of limited processability and low softening temperatures would have to be addressed, but the necessary conductivity is present.

"Metallic" Conducting Polymers

There are four organic polymers with conductivities in excess of $10^0 (\text{ohm cm})^{-1}$: polyacetylene (PA), polyphenylene sulfide (PPS), polypyrrole (PP), and polythiazyl (SN_x). Of these, only SN_x can truly be called an organic metal because in single crystals it possesses (1) a metallic thermopower, (2) high room temperature conductivity, and (3) an inverse temperature dependence of the conductivity.[43] Of all the conducting polymer systems, it alone has the long-range periodic structural order necessary for the valid use of band theory. As expected, the measured conductivities depend strongly on crystal perfection, varying, for single crystals at room temperature, from 250 to 1200 $(\text{ohm cm})^{-1}$. Furthermore, conduction in polycrystalline films is thermally activated and presumably limited by a thermally activated tunneling across grain boundaries. Extensive study has shown that SN_x is not really a one-dimensional conductor as was once thought, but has rather extensive interchain interactions.[44,45]

Although SN_x has been thoroughly investigated in the last two decades, it remains a laboratory curiosity. Application to real devices has been hindered by fabrication difficulties, instability in air at room temperature, brittle mechanical properties,[46] and other factors. Although strong in tensile deformation along the fiber axis, SN_x is very weak in either compression or shear, and it is sensitive to radiation damage. Because mechanical damage degrades electrical properties, this material is unlikely to be useful in rugged environments.

Polyphenylene sulfide (PPS) is perhaps the most processable of the four highly conducting materials. It is a commercially available thermoplastic valued for its relatively high heat distortion temperature, crystallinity, and solvent resistance. Conduction in PPS is realized when it is treated, or doped, with up to 50 mol% of AsF_5, a strong oxidizing (electron-withdrawing) agent.[47,48] As the AsF_5 gas diffuses into the PPS, several physical properties begin to change. The X-ray diffraction pattern becomes increasingly diffuse, indicating that the distribution of intermolecular spacings is broadening. The colorless PPS becomes dark and gets a lustrous metallic sheen. The doped polymer is slightly more brittle than the pristine material and no longer melts below decomposition temperatures.

The processing, mechanical, and thermal properties of virgin PPS make it a very attractive candidate for device applications. However, the extremely rapid increase in conductivity on doping to a 1:1 mole ratio (AsF_5:PPS repeat unit) precludes its use in moderating spacecraft-charging phenomena. The concentration range at which the conductivity increases from 10^{-18} (ohm cm)$^{-1}$ to 10^0 (ohm cm)$^{-1}$ is so narrow, and the distribution of dopant inhomogeneous, so that obtaining a single material with a fixed, homogeneous, moderate level of conductivity would be difficult. A further concern is the stability of the dopant: the PPS:AsF_5 complex is unstable in air and decomposes at 50°C in vacuum.

The conduction mechanism in these highly conducting polymers can be deduced by observing the temperature dependence of the conductivity in these systems. Conductivity in doped PPS shows a clear $T^{-1/2}$ behavior characteristic of tunneling through small barriers between localized conducting domains. This is in marked contrast to the systems previously discussed that displayed a simple, thermally activated, adjacent-neighbor, hopping mechanism. We submit that the density of localized states in doped PPS or other highly conducting polymers is high enough that an effective superlattice is formed and mobility through that periodically spaced superlattice is increased. This leads to tunneling between conducting grains of various sizes, which appears as variable- long-range hopping.

Polypyrrole (PP) is formed by an electrochemical polymerization of pyrrole from solution followed by electrochemical oxidation.[49] The oxidized polymer is a lustrous gold color when doped by approximately one perchlorate ($HClO_4^-$) or BF_4^- group for four pyrrole units. Undoped polymer is stable in air, whereas the doped conducting material is unstable. The doped polymer does not melt or dissolve, and it decomposes on heating to $\sim100^\circ C$. Tensile strengths of the undoped films are relatively high, comparable to those of conventional thermoplastics. Although the thermopower is metallic, the temperature dependence of the conductivity shows a clear $T^{-0.25}$ dependence, as is observed for systems with variable-range hopping between localized states.

We can gain further insight into the effect of structural order on conductivity by examining the maximum conductivities realized in substituted PPs. The highest value reported for the parent polymer is 40 (ohm cm)$^{-1}$; for the N-methyl, 2,3-dimethyl, and 2,3-diphenyl analogs, the conductivity drops to 10^{-3}, 10^1, and 10^{-3} (ohm cm)$^{-1}$, respectively. Although none of these substituents affect the electronic structure of the PP backbone very much, they do affect the intermolecular packing efficiency. The poly(2,3-dimethyl pyrrole) in fact is slightly crystalline, as expected by the relatively high conductivity value reported.

The features that distinguish PP from other semiconducting polymers include (1) a stereoregular backbone that packs efficiently into an ordered, if not crystalline, array; (2) the ability to form a relatively low energy, delocalized, ionized charge state; and (3) the interaction of items 1 and 2 above to form a superlattice of interacting delocalized charge states that lead to the high mobilities inherent in a material with conductivity in excess of 10^0 (ohm cm)$^{-1}$. Evidence for this approach includes the rapid increase in conductivity on doping, the high conductivity itself, the four-to-one mole ratio of repeat unit to dopant, the metallic luster of the doped material, and the effects of substituents on conductivity and crystallinity.

Polyacetylene (PA) also shows a thirteen-order-of-magnitude increase in conductivity on doping to 10 mol%.[50] Significantly, the conductivity is proportional to exp $T^{-0.25}$,[51,52] consistent with the thermally activated variable-range hopping between localized states.[53] This conclusion is further substantiated by measurement of the frequently dependence of the conductivity.[51,52] The exact nature of the localized state is controversial, but most agree that charge is partially delocalized over 10 to 15 repeat units.[54,55]

On a macroscopic scale, polyacetylene films look like aluminum foil, but SEM examination reveals a fiberous structure similar to steel wool. Both theoretical[56] and experimental[57] studies indicate that a considerable amount of structural disorder is present in polyacetylene films, which does not support the application of bandgap or one-dimensional conductor models.

The density of films is highly variable, presumably due to differences in synthetic methods. The density is also consistently lower than the crystallographic density. Strength and modulus are considerably lower than those for Kapton.

The unique localized electrochemical reactions along the PA backbone makes application as a battery material possible, but the serious problems of oxidative instability and repetitive discharges must be solved.[58] As for PP and PS, the lack of thermal stability, the need for corrosive dopants, and the rapid transition, or doping, from insulator to good conductor probably preclude the use of polyacetyene in spacecraft-charging problems.

Conclusions

Polythiazyl has a much greater conductivity than we believe is optimum for application to the spacecraft-charging problem. The dependence of conductivity on dopant concentration for the other three highly conducting materials, polypyrrole, polyacetylene, and polyphenylene sulfide, is so strong that it will probably be very difficult to prepare stable, homogeneous, semiconducting material. None of these materials is particularly stable at elevated temperatures. We estimate that loss of conductivity due to thermal degradation in vacuum will begin at 50° to 100°C. At this temperature the tensile strength and modulus of the polymers will be reduced by about 50% or more. We expect polythiazyl and polyphenylene sulfide to remain fairly brittle at these temperatures, whereas the amorphous content of polyacetylene and polypyrrole is high enough that a ductile mode of failure would probably be observed.

It is doubtful that modifying the backbone structure or substituting polythiazyl or PPS would result in materials with superior properties. However, polyacetylene has literally hundreds of analogs, the conductivities of which are all orders of magnitude lower than the doped parent compound. The loss of structural order on substitution also leads to more soluble, stable, processable, less crystalline materials.

Similarly, substituted PPI are less crystalline and have a lower conductivity than the parent compound. They are not soluble and remain infusible, suggesting that films could be of interest if stability problems could be avoided. One possible approach is to prepare polypyrroles with reactive substituents that could be crosslinked, thermally or with high energy radiation, into a more thermally stable material.

The same problems with conducting PPS, e.g., brittleness and lack of thermal stability, are also problems with structural analogs such as polyphenylene oxide and polymers with pendant groups. The apparent chemical instability of doped PPS at elevated temperatures seems to preclude the successful use of a crosslinked or copolymerized modification.

Products obtained from the pyrolysis of Kapton or polyacrylonitrile appear to be very good candidates in thin film applications. They have the necessary thermal stability, which appears to be the biggest problem in applying polymers to space applications, and can be prepared in conducting forms covering a wide range of conductivity, without the use of dopants. The fused-ring systems that are formed during pyrolysis are very stable to high energy radiation and could well last ten years or more.

The most uncertain feature of these materials is their ultimate mechanical properties. We can be fairly certain that they will become more brittle, but how significant this is will depend on the extent of pyrolysis and specific application. If, for example, we assumed that a "completely" pyrolyzed material would be like graphite, then the worst case would possess a tensile strength of about 10,000 psi and an elongation at break of 2-4%, which would be sufficient for load-bearing films. Furthermore, one can envision pyrolyzing a polymer blend or alloy containing a pyrolyzable component and one of the high thermally stable structural polymers as a reinforcement.

One problem that would have to be addressed is that of progressive pyrolyses and the resultant increases in conductivity in a space environment. The pyrolysis temperatures reported in the literature are generally in excess of $400^{\circ}C$, which is somewhat above that generally encountered in spacecraft, so progressive change would be slow.

Another promising material is polyvinyl carbazole. The polymer combines the thermal stability, radiation resistance, mechanical strength, and conductivity required for many structural applications. It is readily processable, so that it can be used in fibers, films, and castings. Two possible modifications would be of interest. A crosslinked PVK would add another 50° to $100^{\circ}C$ to the glass transition temperature, extending its usefulness to higher temperature extremes. In addition, if a more highly crystalline, stereoregular PVK could be prepared, through the identification of new catalyst systems, then the mechanical propeties and useful temperature would be improved substantially because the glass transition temperature is very dependent on tacticity.

References

1.   J. Hanscomb and J. Calderwood, J. Phys. $\underline{6}$, 1093 (1973).

2.   B. Sharma and P. Pillai, Polymer, $\underline{23}$(1), 17 (1982).

3.   J. Mort and G. Pfister, Poly. Plast. Tech. Eng. $\underline{12}$, 89 (1970).

4.   S. K. Shrivastava and A. P. Srivastava, Polymer $\underline{22}$(6), 765 (1981).

5.   S. Bruck, Polymer $\underline{6}$, 319 (1965).

6.   E. Sacher, J. Macromol. Sci. Phys. $\underline{B4}$(2), 441 (1970).

7.   A. F. Garito and A. J. Heeger, Accnt. Chem. Res. $\underline{7}$, 232 (1974).

8.   W. R. Hertler, J. Org. Chem. $\underline{41}$(8), 1412 (1976).

9.   C. V. Pittman, Y. F. Liang, and M. Ueda, Macramolecules $\underline{12}$(3), 355 (1979).

10. Ibid., $\underline{12}$(4), 541 (1979).

11. M. H. Litt and J. W. Summers, J. Polym. Sci., Chem. Ed. $\underline{11}$, 1339 (1973).

12. J. Ulanski, J. K. Jeszka, and M. Kryszewski, Polym. Plast. Technol. Eng. $\underline{17}$(2), 139 (1981).

13. R. Somoano, S.P.S. Yen, A. Renbaum, J. Polym. Sci. Lett. Ed. $\underline{8}$, 467 (1970).

14. M. Watanabe, A. Tsuchikura, T. Kamiiya, and I. Shinohana, J. Polym. Sci. Lett. Ed. $\underline{19}$, 331 (1981).

15. J. H. Lupinski, K. D. Kopple and J. J. Hertz., J. Polym. Sci., Symp. Ed. $\underline{16}$, 1561 (1967).

16. A. M. Hermann and A. Rembaum, J. Polym. Sci., Symp. Ed. 17, 107 (1967).

17. A. Epstein and B. Wildi, J. Chem. Phys. 32(2), 324 (1960).

18. H. S. Nalwa, J. M. Sinha, and P. Vasudevan, Makromol. Chem. 182, 811 (1981).

19. C. W. Dirk, K. F. Schoch, and T. J. Marks, "Cofacial Assembly of Metallomacrocycles: A Molecular Engineering Approach to Electrically Conductive Polymers," in Conductive Polymers, R. B. Seymour, Ed. (Plenum Press, New York, 1981), p. 209.

20. C. J. Norrell, H. A. Pohl, M. Thomas, and K. D. Berlin, J. Polym. Sci., Phys. Ed. 12, 913 (1974).

21. D. H. Diel, T. Inabe, J. W. Lyding, K. F. Schoch, C. R. Kannewurf, and T. J. Marks, Polym. Prepr. 23(1), 124 (1982).

22. R. S. Nohr, P. Brant, D. Weber, and K. J. Wynne, Polym. Prepr. 23(1), 129 (1982).

23. P. M. Kuznesof, K. J. Wynne, R. S. Nohr, and M. E. Kenney, J. Chem. Soc. Chem. Comm. 121 (1980).

24. R. Liepins, D. Campbell, and C. Walker, Polym. Prepr. 9(1), 765 (1968).

25. H. Teoh, P. D. Metz, and W. G. Wilhelm, "Electrical Conductivity of Pyrolyzed Polyacrylonitrile," BRL report No. 30141 (1981).

26. H. A. Pohl, J. Biol. Phys. 2, 113 (1974).

27. K. Saha, S. C. Abbi, and H. A. Pohl, J. Non-Cryst. Solids 22, 291 (1976).

28. Ibid., 21, 117 (1976).

29. Y. Saski, L. L. Walker, E. L. Hurst, and C. U. Pittman, J. Polym. Sci., Chem. Ed. 11, 1213 (1973).

30. D. O. Cowan, J. Park, C. U. Pittman, Y. Sasaki, T. Mukhjree, and W. A. Diamond, J. Am. Chem. Soc. 94, 5110 (1972).

31. C. U. Pittman and B. Surynavayahan, J. Am. Chem. Soc. 96(26), 7916 (1974).

32. H. Pohl and R. Chartoff, J. Polym. Sci. A2, 2787 (1964).

33. E. L. Thomas, R. J. Farris, and S. L. Hsu, "Mechanical Properties Versus Morphology of Ordered Polymers," AFWAL-TR-80-4045, Vol. III (1982).

34. B. Millaud and C. Strazielle, Polymer 20, 563 (1979).

35. J. Danhaeuser, Makro. Chem. 84, 238 (1965).

36. M. Jozefowicz, L.T. Yu, G. Belorgey, and R. Buvet, J. Polym. Sci. Symp. Ed. 16, 2943 (1967).

37. G. Manecke and J. Kautz, Makro. Chemie. 172(1), 1 (1973).

38. D. M. Carlton, D. K. McCarthy, and R. H. Genz, J. Phys. Chem. 68, 2661 (1964).

39. A. Hankin and A. North, Trans. Far. Soc. 63, 1525 (1967).

40. K. Se, H. Ohnuma, and T. Kotaka, Prog. Polym. Phys. Jap. (1981).

41. G. E. Wnek, J.C.W. Chien, F. E. Karasz, and C. Peter Lillya, Polymer 20, 1441 (1979).

42. P. Cukor, J. I. Krugler, M. F. Rubner, Makro Chemie. 182, 165 (1981).

43. C. H. Hsu and M. M. Labes, J. Chem. Phys. 61, 4640 (1974).

44. R. H. Dee, J. F. Carolan, B. G. Turrell, and R. L. Green, Phys. Rev. B22(1), 174 (1980).

45. P. M. Grant and I. P. Batra, Synth. Metals 1, 193 (1980).

46. R. H. Baughman, R. H. Baughman, P. A. Apgar, R. R. Chance, A. G. MacDairmid, and A. F. Garito, J. Chem. Phys. 66(2), 401 (1977).

47. T. C. Clark, K. K. Kanazawa, V. Y. Lee, J. F. Raybolt, J. R. Renolds, and G. B. Street, J. Polym. Sci., Phys. Ed. 20(1), 117 (1982).

48. R. R. Chance, L. W. Shacklette, H. Eckhardt, J. M. Sowa, R. L. Elsenbaumer, D. M. Ivory, G. G. Miller, and R. H. Baughman, "Conducting Complexes of a Processible Polymer: Poly(p-phenylene sulfide)," in Conductive Polymers, R. B. Seymor Ed. (Plenum Press, New York, 1981), p. 125.

49. K. K. Kanazawa, A. F. Diaz, W. D. Gill, P. M. Grant, G. B. Street, G. P. Gardini, and J. F. Kwak, Synth. Metals 1, 329 (1980).

50. Y. W. Park, A. J. Heeger, M. A. Druy, and A. G. MacDaiarmid, J. Chem. Phys. 73(2), 946 (1980).

51. A. J. Epstein, H. W. Gibson, P. M. Chaikin, W. G. Clarke, and G. Grumer, Phys. Rev. Lett. 45(21), 1730 (1980).

52. Ibid., 47(21), 1549 (1981).

53. S. Kivelson, Phys. Rev. Lett. 46(20), 1344 (1981).

54. Y. Tomkiewicz, T. D. Schultz, H. D. Brom, T. C. Clarke, and G. B. Street, Phys. Rev. Lett. 43(20), 1532 (1979).

55. S. Ikehata, J. Kaufer, T. Woerner, A. Pron, M. A. Druy, A. Sivak, A. J. Heeger, and A. G. MacDairmid, Phys. Rev. Lett. 45(13), 1123 (1980).

56. D. Vanderbilt and E. J. Mele, Phys. Rev. B22(8), 3939 (1980).

57. V. Enkelmann, W. Muller, and G. Wegner, Synth. Metals 1, 185 (1980).

58. Y. W. Park, M. A. Druy, C. K. Chiang, A. G. MacDiarmid, A. J. Heeger, H. Shirikawa, and S. Ikeda, J. Polym. Sci., Lett. Ed. 17, 195 (1979).

7. Conductivity in Organic Polymers

An initial reading of the literature concerning electroactive polymers seems to indicate a different mechanism responsible for conductivity for each material studied. While this confusion is partly due to the complexity of conduction mechanisms in polymers, it is certainly aggravated by the following three observations.

First, most researchers in the field select a single material or type of material as the best material and promote its maximum achievable conductivity. Second, few of those active in the field of conducting polymers are formally trained in polymer science or have much experience in the area. Collaboration between synthetic chemists and solid state physicists is excellent, but, as we will show, an improved understanding of polymer science (particularly morphological and structural phenomena) can help end much of the confusion in the literature. Third, published reviews in the field,[1-5] with perhaps one exception,[6] are little more than lists of materials and their properties. Almost no attempt is made to systematize data according to either structure or mechanism.

The primary purpose of this project was to prepare a comprehensive theory of conductivity in polymers. Secondary purposes included selecting contemporary materials and designing new ones for very specific applications. Thus, it required both a broader perspective of materials and a more unified interpretation of the physical process involved in conduction than previous reports. A more explicit recognition of the varying degrees of structural order in polymers and the nature of the charge carriers that can operate within the limitations of that order can contribute significantly to understanding the wealth of data reported in the literature.

Structural order is important to conductivity mechanisms. Consider the two structural extremes for which good theoretical models exist: inorganic crystalline lattices[7] and amorphous semiconductors.[8] Because of the long-range periodic order present in crystalline lattices, the electronic configuration of individual atoms can be easily combined to form long-range delocalized bands of discrete energies. The population of each energy level can be calculated and the level of conductivity observed is due to the relative population of valence and conduction bands. Insulation can be understood in terms of the energy gap between valence and conduction bands. Likewise, semiconduction is relatively well explained by the thermal activation of electrons from valence to conduction band. Dopant molecules with different electronic populations can contribute to the conductivity either by distorting the lattice energies (interstitially) or by providing a conduction band electron (valence band hole) by occupying a lattice site (substitution).

At the other structural extreme are the amorphous semiconductors. The validity of the recently developed theoretical models has been demonstrated in many materials. These materials are truly disordered. Conduction occurs because the band edges are smeared out, leading to what can more properly be called a mobility gap rather than an energy gap. Charge carriers are now highly localized at sites of structural disorder. The energy difference between these sites is relatively small, compared with the surrounding media, so conduction is limited by mobility rather than number of carriers. Localized charges therefore hop instantaneously, through a weak thermally activated process, from site to site.

Few, if any, polymeric materials realize these extremes. On one hand the nature of a polymeric material requires that a correlation exists between the position of one repeat unit and the position of its two covalently bound nearest neighbors. On the other hand, the entropic and kinetic restraints to forming evenly spaced microscopic lattices out of highly entangled chains thousands of angstroms long prevent the formation of crystalline arrays necessary for the application of band theory.

The remainder of this section consists of three parts. The first summarizes the nature of carriers responsible for conduction and the types of structural order associated with each carrier. The second discusses the mechanisms responsible for the transport of these carriers and the dependence of that mechanism on temperature, pressure, dopant and structural order. The third summarizes some of the problems that are unresolved by these models and that a more generalized model should treat.

Charge Carriers

One way to begin to understand the profusion of data reported on conduction in polymeric systems is to classify materials according to the carriers responsible for conduction. Although obtaining the experimental data necessary to unambiguously identify the carriers involved can be very difficult, the data are finite and they are generally associated with different degrees of structural order.

The first species are ionic carriers.[1] These are distinct, ionically charged chemical entities: hydrogen or hydronium ions, acids, bases, salts, metallic impurities due to residual catalyst, or species intended to be electronic dopants but which end up as mobile ionic species. Ionic conduction occurs throughout the entire structural range, from polycrystalline to amorphous, and requires actual mass transport. It is extremely sensitive to pressure and humidity. Ionic conductivity is always possible when readily ionizable species are present in the polymer as possible degradation products, or when purity is uncertain. In any experimental investigation the question of possible ionic conductivity should be immediately addressed.

The second species or state that we must consider are localized electronic excited states.[9] The key word is "localized." These energy states, although they can contribute to electron spin resonance (ESR) signals if their spins are unpaired, are localized on a single molecular residue or repeat unit and hence cannot participate in conduction.

The third most energetic species is the intramolecular exciton.[10] An exciton is a neutral excitation consisting of an electron and the positive species it leaves behind. Transport of this "particle" (coupled positive and negative charge) in itself does not lend to conduction. The transport occurs with the concerted cooperative polarization of the lattice, both for a periodic atomic lattice and a molecular lattice found in organic single crystals. Both inter- and intramolecular excitons are possible in molecular lattices, requiring increasingly cooperative lattice polarizations. This neutral particle contributes to bulk conduction when the neutral species is separated or ionized by an applied electric field.

Stable, bound excitons require some local order: a regular, polarizable environment prevents the charged species from recombining. However, the extent of this spatial periodicity need not be great. Because the intermolecular interaction is coulombic, it is readily screened within a few molecular lengths in most polymers. The polaron, a carrier related to the exciton, is an exciton coupled to periodic lattice vibrations.[11] This of course requires an extensive long-range periodic lattice structure to operate. Excitons and polarons originated in band gap theories to explain the existance of stable states with energies between those of the valence and conduction band edges.

The fourth species is migratory localized ionic states,[12,13] particularly anions. Although these states are truly ionic, the transport mechanism responsible for conduction is electronic. The species involved is usually a discrete molecule or molecular residue with a high electron affinity. Anions may be formed through electron transfer that is photoinitiated, chemically driven as in a donor:acceptor complex, or perhaps caused by an electrochemical redox reaction. In systems where these carriers are responsible for conduction, very little intermolecular structure is observed, and transport occurs through a random hopping mechanism. Essentially, the only structure necessary is the extremely local cofacial orientation of donor (or photoinitiated donor or electrochemically oxidized species) to the acceptor.

A fifth and relatively new carrier, whose existence and significance is hotly contested, is the "soliton." The name is taken from a hypothetical particle associated with the solution of a complex set of differential equations that apply to a variety of physical phenomena.[14] This species can possess a positive, negative, or neutral charge, and it is believed to be responsible for conductivity in quasi-one-dimensional models for polyacetylene. In polymers that can be modeled as one dimensional, the soliton is a defect or kink in the phase of the alternating carbon-carbon double bonds along the chain backbone.[15-16] Some doubt the existence of the soliton, based on fundamental and experimental grounds. Others are willing to concede its existence, but question the need to invoke a unique particle to explain the properties of a single material.[17]

The sixth and final carrier species is the electron or hole carrier associated with band theory.[7] These species are free to migrate with an applied field, moderated by a periodic fluctuations in the band energy due to the thermal vibration of the lattice.

The charge carrier responsible for conduction in band theories of inorganic amorphous semiconductors are again electrons, or the absence of an electron. They tend to be spatially localized at the site of structural disorder in amorphous glasses. These sites in amorphous silicon are believed to be "dangling bonds," a site on a given silicon atom that is available for covalent bonding but is not bound because there is no neighbor at the correct distance and at the correct orientation.[18] The disorder in these systems is even greater than for amorphous polymer glasses because in polymers the position of at least the adjacent repeat unit or atom is always known.

Before concluding this section, we summarize the scale on which order may be present, e.g., the dimensions at which the probability density function reaches a maximum. The crystalline atomic lattices to which band theory applies possess regular periodic structures over distances of 10^5 to 10^8 Å. In amorphous inorganic semiconductors all possible correlation between the position of adjacent atoms has decayed to zero within a few atomic diameters, say 5 Å.

In organic polymers the situation is strikingly different. The atoms that make up the chain are covalently bound with well-known bond angles and bond distances. Thus there is, in even the most amorphous polymer, a finite degree of correlation between the position of one chain atom and two or three on either side. For rodlike polymers, even though they are not crystalline in the normal sense, this chainwise periodic structure exists over lengths of nearly 5,000 Å.

The periodic structure of organic polymers discussed above, due entirely to the covalently bound chain structure, exists independently of normal polymer crystallinity, which involves three-dimensional periodicity. With the exception of two materials that can be polymerized directly from monomer single crystals to polymeric single crystals, the long-range intermolecular periodicity required to form energy bonds is not in polymers or is on such a small scale that it does not contribute to the bulk electrical properties.

Why are band gap theories applied to polymers if they clearly do not possess the structural order from which energy bonds are derived? There is a popular misconception concerning electron delocalization along a conjugated polymer backbone: it is mistakenly believed that overlapping atomic orbitals in a polymer backbone can combine to form highly extended, delocalized molecular orbitals in the same way that atomic orbitals in an ionic crystal form delocalized energy bands. The popular misconception is that electrons race back and forth along the backbone in very low (binding) energy molecular orbitals. Thus one would predict high conductivities, very low energy electronic (optical) transitions, large dielectric constants, and large electron spin resonance signals. But for the majority of polyconjugated systems none of these properties are observed. Although periodic structure of variable lengths along the chain may be important to conduction it does not lead to band type conduction because it lacks the necessary intermolecular structure.

To summarize, we have reviewed the types of charge carriers that may be involved in conduction in polymers and their physical nature. We have introduced the new idea that the scale on which structured order is present can influence the type of carriers active and hence better rationalize the conductivity data in the literature.

Transport Mechanisms

With several types of charge carriers known to contribute to
conductivity in polymers, it is not surprising to find references to a
great many mechanisms for charge transport in the literature and a
confusing array of temperature, pressure, concentration, and applied
field dependencies. We first summarize these mechanisms and their
dependence on external variables and then, as in the previous section,
relate the mechanism responsible for charge transport in terms of
structural order.

Conductivity is equal to the product of the carrier mobility, μ,
its charge, q, and the number of carriers, n.[6] More than one type (i) of
charge can be present in a given material, and therefore

$$\sigma = \Sigma_i \mu_i e_i n_i \qquad (7\text{-}1)$$

Depending on the specific mechanism involved, all three variables on the
right-hand side of equation (7-1) are dependent on the environment and
can be modified to some degree.

There are five relatively simple conduction mechanisms and a few
complex ones, which involve either combinations of the first five or
variations thereon.

Ionic Conductivity

All three terms of equation (7-1) can be used to modify ionic
conductivity. Charges per carrier of as much as ± 3 have been observed,
and the number of carriers is extremely sensitive to prior treatment.
Three basic sources of ions are available. The first source is ionized
particles normally bound to the polymer backbone, e.g., pendant
carboxylic or sulfonic acid groups, amines, and amide groups. This
dissociation is usually temperature dependent, so both the number of
carriers and their mobility may be changing at the same time with

temperature. The second most frequently encountered source of carriers is inadvertent impurities. These include catalyst residues, degradation products, and dopants that are intended to enhance electronic conductivity. The third source, and probably the most troublesome, is water. Almost all polymers will adsorb 0.1 to 1% water, which by itself or with ionizable groups or impurities can greatly enhance the observed conductivity.

Because the charge transport associated with ionic conductivity requires actual mass transport, it has certain unique features.[1] It is thermally activated, but increasing pressure reduces the mobility of carrier ions. As conduction in these systems requires mass transport, and hence open channels in the material, it is extremely sensitive to variations in fabrication. Conductivity is given by equation (7-2):

$$\sigma_{ionic} = {}_i\Sigma_{species} \ (K_i n_i)^{1/2} \ q_i \mu_i \ \exp\left(\frac{\Delta w_i}{2\epsilon kT}\right) \tag{7-2}$$

In equation (7-2), n_i is the number of ionizable species, K_i is the dissociation constant for these species, μ_i the mobility of change q_i and ϵ is the relative dielectric permitivity. The presence of water, with a dielectric constant 25 to 50 times larger than most polymers, dramatically affects conductivity.

The exact form of the mobility μ depends further on the specific structure of the material.[2] A frequently used model assumes that an ion passes with a characteristic frequency ν across a potential energy barrier Δw between localized sites of size α in the polymer matrix. An applied field modifies the local potential, and the mean drift velocity v in the field E direction is given by equation (7-3) and the current density flowing through a sample in equation (7-4):

$$v = \exp\left(-\frac{Ea}{kT}\right)\left[2 \ \sinh\left(\frac{e\alpha E}{2kT}\right)\right] \tag{7-3}$$

$$j \propto \exp\left(\frac{-v^*}{v_f} - \frac{\Delta w}{2\epsilon kT} - \frac{Ea}{kT}\right)\left[\sinh\left(\frac{e\alpha E}{2kT}\right)\right] \tag{7-4}$$

Increased pressure reduces the free volume available V_f for transport relative to the minimum volume V^* necessary for transport. Below the polymer glass transition temperature, V_f is only weakly sensitive to either temperature or pressure; however, above T_g it rapidly increases with increasing temperature and rapidly decreases with increasing pressure.[19]

Since charge transport is due to a finite number of carriers initially present, the current density j may be observed to decrease rapidly as several coulombs are passed through the sample and the supply of carriers is depleted. This is, of course, a DC effect and can be removed by reversing polarity. Generally crystallinity reduces ionic mobility due to the increased density of the polymer matrix. In at least one system, the polyamides (nylons), stretch-induced biaxial orientation leads to anisotropic conduction because of spatial ordering of the amide repeat groups.[20]

Band-Type Conduction

In band theory, atomic orbitals combine to form a spatially extended, delocalized energy band. The conductivity then depends on the relative population of each band and the energy difference between bands. Charge transport can occur only in partially occupied bands. For example, in metals the valence and conduction bands merge and there are more energetically equivalent orbitals than electrons to occupy them. In insulators the valence band is full, and the conduction band is empty, and the energy gap between them is considerably greater than the thermal energy required to promote a valence electron into the conduction band. Even when the necessary energy is present, either in very high temperatures or high energy radiation, the conduction electron:valence hole pair will not contribute to conductivity unless a fairly high electric field is applied. Otherwise the columbic attraction between the electron hole pair leads to recombination.

When the energy gap between valence and conduction bands is comparable to kT, there is enough thermal energy to promote electrons into the conduction band, and intrinsic semiconduction occurs. The distribution of electrons f(E) between states of various energies E is given by equation (7-5)

$$f(E) = \frac{1}{1 + \exp(E - E_f)/kT} \tag{7-5}$$

where E_f is the fermi energy corresponding to the energy of a state, and the average probability of the state being occupied is 1/2.[1]

The number of conduction electrons n (equal to the number valence holes, p) is then given by equation (7-6) where m^* is the effective mass, h is Planck's constant, and Ea is the energy gap.

$$n = p = 2(\frac{2\pi m^* kT}{h^2})^{3/2} \exp(Ea/2kT) \tag{7-6}$$

For metals the overlap between bands is large and there are many carriers. The conductivity is insensitive to pressure and decreases with increasing temperature due to enhanced electron-electron and electron-phonon scattering. However, in semiconductors increased pressure lowers the energy gap between bands and increased temperature increases the population of the conduction band; both raise the overall conductivity. Mobilities in semiconductors are generally unaffected by pressure and weakly diminished with increased temperature through the same mechanism found in metals. Mobilities are generally related to the width of the conduction band, which is in turn due to the extent of atomic orbital overlap. Band theory begins to break down when the extent of orbital overlap falls to where the mobility is under about 1 cm^2/V sec. At this point the mean free path of an electron between scattering collisions is comparable to the atomic lattice spacing or, alternatively, there is very little electron delocalization and the band degenerates into a bound excited state (treated in the following section).

Defects or impurities can promote electrons into the conduction band. For example, replacing an atom on the lattice with a different chemical species with a different electron density can create either a valence hole or an extra conduction band electron. Alternatively, an interstitial atom can distort the local band energies and modify the thermal equilibrium population of the bands. Very small concentrations of impurity atoms, or deliberately included dopants, can significantly change the valence band population. Temperature and pressure sensitivity of extrinsic semiconductors is similar to that discussed for intrinsic conduction.

Hopping Transport Between Localized Electronic States

As the extent of atomic, or molecular, orbital overlap decreases, mobility is lowered and electronic states become increasingly localized. In the band gap models, the energy of the band edges smears out, and the density of states indicates new energy states in what once was the band gap. Although there may be a continuum of positive, finite, populated energy states in what was originally the band (energy) gap, these states are now so localized that conduction is severely mobility-limited, and the space between valence and conduction bands becomes a mobility gap rather than an energy gap. Charge transport can now only occur through the thermally activated hopping across potential energy barriers between localized states.

A first-order approach for calculating the mobility of a hopping particle is given in equation (7-7) where D is an effective diffusion coefficient given by equation (7-8) where ν is the hop frequency and L the mean hop length.[21]

$$\mu = eD/kT \qquad (7-7)$$

$$D = (1/2)\,\nu\,L^2 \qquad (7-8)$$

For diffusion in a matrix of identical states spaced evenly with a

single-valued interstate potential energy barrier, diffusion is a
thermally activated process, and the mobility due to normal hopping is
given by equation (7-9).

$$\mu = \mu_o \, \exp(-E_a^{hopping}/kT) \qquad (7\text{-}9)$$

Of course real systems contain a distribution of energy states, and a
distribution of hop lengths are encountered. The mobility for such a
variable-range hopping transport is then

$$\mu = \mu_o \, \exp[-(T_o/T)^{1/4}] \qquad (7\text{-}10)$$

where T_o and μ_o are experimentally determined constants.

Hopping models also account for the transport of localized
electronic states in lattices undergoing quantized thermal vibrations
(electron-phonon coupling). The quantized energy needed to overcome the
potential barrier and equation (7-9) becomes

$$\mu = \mu_o \, \exp(-E_a^{polar}/kT) \qquad (7\text{-}11)$$

where $E_a^{polar} \leq E_a^{hopping}$.

Excitonic Conduction

The dominant mechanism responsible for conduction in the majority
of polymers is due to the field-induced ionization of neutral
excitons.[10] This is also a hopping process, but both the generation and
mobility of carriers is influenced by the applied field. The
conductivity at high fields E is therefore given by equation (7-12) as
the product of a constant, the carrier concentration term (in brackets)
and a mobility term (in parenthesis) where β_F is an ionization constant
for the neutral exciton and ε the relative dielectric permittivity.

$$\sigma = \sigma_o \left[\frac{2 + \cosh(\beta_F E^{1/2}/kT)}{3} \right] \left(\frac{2kT}{\varepsilon E} \, \sinh \frac{\varepsilon E}{2kT} \right) \qquad (7\text{-}12)$$

A similar phenomenological expression is found in the mobility of photogenerated radical anions in molecularly doped polymers and amorphous glasses.[13] Charge transport is again due to hopping, is thermally activated, and is dependent on the applied field as indicated by equation (7-13) where ρ is the hop length.

$$\mu = (L)^{-1/a} \left[\sinh\left(\frac{e\rho E}{2kT}\right)^{1/a} \right] \left[\exp(-E_a/kT) \right] \qquad (7-13)$$

In these systems carrier diffusion is highly dispersive, in contrast to the more frequently encountered Gaussian distribution. The distribution of carriers hopping times is given by equation (7-14a) rather than the Gaussian factor in equation (7-14b) In Equation (7-14a), $a = a_0/(1 - T/T_0)$; and a and T_0 are empirical constants with $o < a < 1$. These relationships can perhaps be considered the variable-frequency analogs to the variable-range hopping described in equation (7-10).

$$\phi(t) \sim t^{-(1+a)} \qquad (7-14a)$$

$$\phi(t) \sim \exp^{-(t/\tau)} \qquad (7-14b)$$

Quantum Mechanical Tunneling Between Metallic Domains

The final conduction mechanism involves charge transport in a material that may be thought of as polycrystalline. Here we observe metallic or semiconducting domains separated by an orientation boundary or amorphous region across which mobility is due to quantum mechanical tunneling.[22] The conductivity for this type of morphological structure is given in equation (7-15):

$$\sigma = \sigma_0 \exp[-(T_0/T)^{1/2}] \qquad (7-15)$$

The single feature most useful in connecting these modes of transport is their temperature dependence. In band gap metals log σ is inversely proportional to temperature due to increased scattering. In band gap semiconductors log σ is proportional to temperature because carriers must be thermally excited into the conduction band. A similar dependence is observed for the nearest neighbor hopping systems regardless of whether or not the molecular structure of the charge carriers is well defined or diffuse. Likewise, variable-range hopping and tunneling transport can be distinguished by a fractional temperature exponent, whereas excitonic conduction in most insulating polymers is a complicated temperature and electric-field-dependent process. These distinctions become useful when one examines specific materials and attempts to understand and predict electrical properties.

Unresolved Problems

The previous sections summarize several models that describe conductivity in systems of different structures and properties. With this diversity of carriers and transport mechanisms, one might think that electrical conductivity in organic polymers would be well understood. However, this is not the case. In this section, we will briefly review some of the discrepancies between these straightforward models and the actual, observed, experimental properties.

The general approach that "molecular architects" have used is based on the idea that in order to obtain electron transport through an organic media, one builds an extended π-electron molecular orbital. This type of structure is indeed present in highly conducting polythiazyl and TTF-TCNQ complexes, but it is also present in numerous insulating or semiconducting polymers.

Even if complete intramolecular electron delocalization could be obtained, as in a benzene ring where the circulation of electrons in radical ions generates a measurable magnetic movement, these are still molecules with characteristic dimensions of 100 to 1000 Å. In the absence of molecular vibrations, the mean free path in such a system is at best about 100 Å. When the three-dimensional structure of polymer molecules is taken into account, this distance becomes considerably shorter than that observed for crystalline band gap conductors.

Given intramolecular delocalization, many intermolecular transfers must take place if charge is to be transported over finite distances. Many more systems exhibit metallic thermoelectric coefficients and larger ESR spin concentrations than show significant electrical conductivity. This is particularly important in spacecraft charging because extremely large, efficient, intermolecular charge transport is essential if the many secondary electrons created on dielectric surfaces are to be drawn away before being discharged. Because all methods currently used to calculate conductivity are based on the electronic structure of a single repeat unit and ignore the need for intermolecular transfer and finite mean-free-path lengths, considerable opportunities exist for theoretical and experimental development.

Finally, whatever their electrical properties, polymers for spacecraft charging must be processable in addition to processing mechanical integrity and environmental stability. The highly conducting polymers are not processable by conventional techniques. Neither are many of the more interesting semiconducting polymers. If a solution to the problem of spacecraft charging is to be found in the improved design and selection of materials, new processing schemes must be developed.

References

1.   A. R. Blythe, Electrical Properties of Polymers, Cambridge Solid State Science Series (Cambridge, England, 1979).

2.   A. Rembaum et al., "Polymeric Semiconductors," in Progress in Dielectrics, Vol. 6 (Academic Press, New York, 1965).

3.   E. Goodings, Chem. Soc. Review 5, 95(1976).

4.   T. Fabish, Critical Reviews in Solid State and Material Science 8(4), 383 (1979).

5.   H. Block, Adv. Polym. Sci., 33, 93 (1979).

6.   C. Duke and H. Gibson, "Conductive Polymers," Encyclopedia of Chemical Technology, Vol. 18, 3rd Ed. (J. Wiley & Sons, New York, 1982).

7.   J. Ziman, Principles of the Theory of Solids (Cambridge University

Press, New York, 1964).

8.   D. Adler, Scientific American, 36 May (1977).

9.   H. Pohl, J. Polym. Sci. Sym. Ed. 17, 13 (1967).

10.   J. Frenkel, Phys. Rev. 37(17), 1226 (1931).

11.   D. Emin, Phys. Today 35(6), 34 (1982).

12.   J. Vlanski et al., Poly. Plast. Tech. Eng. 17(2), 139 (1982).

13.   J. Mort and G. Pfister, Poly. Plast. Tech. Eng. 12, 89 (1970).

14.   A. Scott et al., Proceedings of the IEEE 61(10), 1443 (1973).

15.   W. Su et al., Phys. Rev. Lett. 42(25), 1698 (1979).

16.   G. Street et al., Phys. Rev. Lett. 43(20), 1532 (1979).

17.   A. MacDiarmid et al., J. Chem. Phys. 73(2), 946 (1980).

18.   W. Mott, Contemp. Phys. 10(2), 125 (1969).

19.   K. Saha et al., J. Noncryst. Solids 21, 117 (1976)

20.   E. Sacher, J. Macromol. Sci., Phys. B19(1), 131 (1981).

21.   H. Pohl, J. Biol. Phys. 2, 113 (1974).

22.   C. Chiang et al., Solid State Communications 18, 1451 (1976).

8. Generalized Theory of Conductivity in Organic Polymers

Our survey of materials and mechanisms indicates that there is no one model or theory that can qualitatively predict conductivity. The range of structures and properties are too great to be accounted for with a single highly developed model. What experimental facts must a unified theory encompass and what would we like such a theory to predict? First, we must explain why so many modes of conduction are believed to operate in organic polymers. If the materials could be divided into clear-cut categories, e.g., variable-range hopping, thermally activated semiconductors, and metallic conductivity, this task would be much simpler. However, a few mole percent of a dopant can dramatically change the mechanism of conduction and raise the conductivity, as well as the mobility, from insulator to metallic levels, and such a phenomenon is unparalled in the field of solid state physics. In addition, the extent of structural order necessary to form delocalized conduction bands is simply not present in any of the highly conducting polymers. In short, such a model must account for a wide range of properties within the various degrees of structural order encountered. To be really useful, this model must also allow one to evaluate the conductivity of proposed chemical structures, preferably by separately estimating the number, charge, and mobility of charge carriers.

The Conceptual Model

Our model is based on the notion that the fundamental structural unit is a spatially localized, but electronically diffuse, charge state. It is diffuse in the sense that a given electron will be delocalized throughout this state, which may contain many repeat units on more than one polymer molecule. It will still be essentially microscopic and localized with respect to bulk conductivity. The actual size of this species will depend on the degree of local supermolecular structure present in the polymer. As we indicated earlier, structural order refers not only to lateral order associated with the packing of chains but also to the degree of longitudinal order along the chain backbone with respect to a vector passing through the local backbone.

This localized charge state is similar to that involved in hopping transport. The main difference is that we envision a species whose size and electronic energy can vary considerably depending on the local chemical composition and local structural order.

However, there remains the need to explain how and why metallic conductivity, and the associated requirements of high mobility and band structures, can exist and to explain the rapid, continuous, transition from semiconductor to metal within a range of a few percent of a dopant. Alternatively, we need to explain how phenomena as diverse as variable-range hopping, thermally activated semiconduction, and metallic conduction can all occur in the same material over a narrow dopant range. We propose that these static, but delocalized, states can interact to form a superlattice, analogous to atomic orbitals forming extended molecular orbitals and bands. Therefore, the three factors that interact to determine the overall conductivity are the number, size, and energy of the localized states.

Generally, the relative energy of the states in a given material will be constant, so let us first examine how the number and size combine to determine the conductivity in a material. When relatively few states exist, their distribution in space will be random and noninteracting. In such a situation charge transport can proceed only through a hopping mechanism. The important quantities that need to be calculated are the distributions of hop lengths and their lifetimes.

As the number of charge-carrying states is increased, the mean distance between states is decreased, and eventually the states will begin to overlap, as do atomic orbitals. Extended bands will begin to form, but at this intermediate concentration, which is related to the effective size of the charge state, they will be spatially small and different in size. Mathematically, this situation is related to percolation in three dimensions. The extent of delocalization is not large enough for bulk conduction to be modeled by metallic bands, but the energy of the delocalized states is changing and hence so is the probability of hopping from one site to another. As in percolation, the overall mobility increases rapidly as the number and size of delocalized states grows.

When a critically large number of localized charge states have been created, or populated, we envision that a superlattice of electronically diffuse but spatially localized charge states is formed. The concentration at which this supperlattice forms is dependent on the effective size of the localized states, which in turn is determined by the chemical constitution and morphological order. Intermediate thermally activated semiconductor states exist in the narrow concentration regime that has lattice vacancies and for lattices in which limited site develop creates a band-gap-type situation.

The same type of temperature dependencies discussed in Section 7 would be observed. A further consideration needs to be added for organic polymers. In atomic or molecular crystals thermal vibrations are generally small in amplitude and symmetrically distributed about the center of mass. However, in polymers there is considerably more freedom of motion. Hence, a considerably greater range of inter- and intramolecular overlap is possible. This overlap is time dependent, although slow with respect to electron motion, and will increase substantially at temperatures above the glass transition temperature.

Dopants interact with this model in two ways: they can create or populate localized states, and they can vary the depth of the potential energy well associated with each state. Part of the experimental evidence supporting this model is the relatively high quantities of dopant, usually a 1:1 mole ratio with respect to the polymer repeat unit, necessary for the development of metallic conductivity. In band gap semiconductors conductivity increases in proportion to the dope concentration as new carriers are generated. However, in polymers the transition is extremely sharp. In the following section we examine some

quantitative relationships that used to define our model more
specifically.

Quantitative Relationships

In this section we consider first the characteristics of localized
states; second, how the interaction of localized states affects charge
transport; and third, what these mean to the observed macroscopic
conductivity.

Simple Models for a Localized Charge State: The Particle in a Box

The way to begin to understand charge transport in solids is
through the so-called "particle in a box" calculations. One associates
the probability of an electron escaping from a molecule or other
localized state with the escape of a particle from a potential energy
well. In a one-dimensional well of width el, the energies available to
a particle of mass m are, of course, quantized

$$E = \frac{h^2 n^2}{8\, m\, el^2} \qquad (8\text{-}1)$$

and depend strongly on the number of particles present (since only two
particles can share the same quantum number n) and on the size of the
well.[1] The probability that a particle will tunnel out of the well
depends on the particle energy E and the well height V_o. The basic
result from the simple well is that there is a finite, if small,
probability that the particle can be found outside the well where no
stationary state exists.

A somewhat more useful question to ask is the probability of a
particle escaping through a barrier into a low potential continuum. In
this instance, the probability of escape is finite and stable states
outside the well exist. Introducing the barrier width (b - a) defined
in Figure 8-1, the transmission coefficient for escape to an external
stationary state in three dimensions is given by[1]

$$T = \cos^{-1}\left[\sqrt{a/b} - \sqrt{a^2/b^2}\right] \ \exp\left(-2\sqrt{2m(V_o - E)/h^2}\right) \qquad (8\text{-}2)$$

The inverse cosine function is plotted as a function of a/b in Figure 8-1. This indicates that, once the well width equals the barrier width (a/b = 0.5), the transmission coefficient has reached 87% of its theoretical maximum. Increasing the well width by a factor of 10 increases the transmission coefficient by only about 10%. Thus, for later calculations of the charge mobility, we anticipate that localized charge state volume will have little to do with the probability of an electron tunneling out of a given localized state.

Macroscopic Transport in Nearest Neighbor Hopping Systems

Because our basic approach to understanding the conduction process in organic polymers is based on the role of localized electronic states, we need to summarize how the local structure of those states affect macroscopic charge transport. Nearest neighbor hopping can be treated as a diffusion process with the mobility μ given as

$$\mu = |e|D/kT \qquad (8\text{-}3)$$

where e is the charge on an electron and D is the diffusion coefficient. In a completely general sense, D is given as

$$D = \frac{1}{2}\left(\frac{N_x}{3} + \frac{N_y}{3} + \frac{N_z}{3}\right)\left(\frac{1_x}{3} + \frac{1_y}{3} + \frac{1_z}{3}\right)^2 \qquad (8\text{-}4)$$

where N_i and 1_i are the hop frequency and hop length in the ith direction, respectively. For an isotropic system, equation (8-4) reduces to the familiar result $D = 1/2 \ N1^2$. Thus, for a nearest neighbor hopping mechanism the charge mobility and the conductivity can be obtained through calculation or estimation of the hop frequency and hop length.

With respect to the hop length 1, we can readily establish approximate values for various physical systems. In molecular crystals (e.g., anthracene or naphthalene), 1 is equal to the intermolecular distance, approximately 1.4 Å. In molecularly doped systems (e.g.,

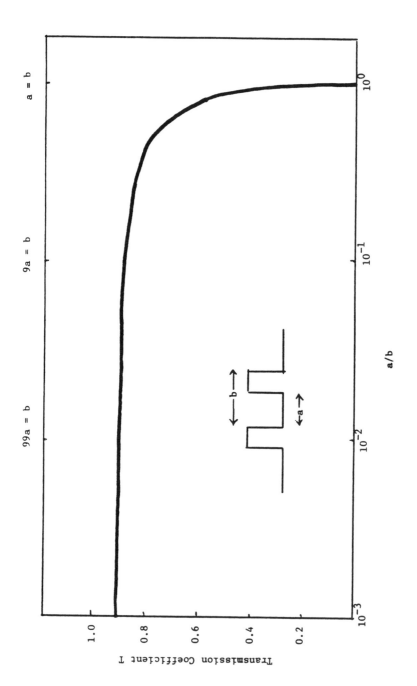

FIGURE 8-1   TRANSMISSION COEFFICIENT FOR ELECTRON ESCAPE FROM A POTENTIAL WELL AS A FUNCTION OF BARRIER TO WELL WIDTH a/b

triphenyl amine in polycarbonate), an average hop length between triphenylamine molecules can be calculated based on the assumption of a homogeneous distribution of segments and geometrical arguments. At a fixed volume fraction of dopant, ϕ_D, the average hop length between dopants becomes $1 \sim 2(3\phi_D/4)^{1/3}$. Charge mobility is therefore found to be proportional to $(\phi_D)^{2/3}$.

Extending this simple model for estimating 1 becomes more difficult for real systems because the identity of the inter- or intramolecular localized electronic state is unclear. However, we show later how the use of percolation methods allows an estimate to be made with reference to molecular structure.

Estimating the hop frequency N is more difficult, but again in a general sense, one can express N as the product of the number of times an electron tries to escape from its localized state times the fraction of times it succeeds. Pohl and his collegues[2] performed this evaluation based on a simple particle in a box model and arrived at calculated mobilities very close to those experimentally observed from anthracene and other molecular crystals.

Macroscopic Transport Involving Anisotropic Nearest Neighbor Hopping

Before continuing to explore how localized electronic states can interact leading to metallic mobilities, let us briefly examine transport in an anisotropic system. Our objective is to compare the experimental conduction anisotropies, observed in so-called "quasi-one-dimensional conductors" that are supposed to have very long-range electron delocalization along one molecular axis, with transport anisotropies calculated using a biased random walk on an anisotropic lattice. Although this model is not exact, particularly because it ignores interparticle interactions, it does illustrate how large an effect the bias of an applied field and an anisotropic hop length have on transport anisotropy.

In this model we have a reflecting boundary at position X and an absorbing barrier at X = a. The total hop rate is fixed at unity, and the probabilities of hops right and left are given by p and q, respectively. Thus, for an isotropic system, p = q and p + q = 1/3; that is, 1/3 of all hops are along each x, y, and z axis, respectively, and there is an equal probability of hops toward or away from the absorbing barrier. By varying the sum p + q, we can increase the fraction of hops along the x axis (e.g., along an axis of increased or preferred orientation) relative to the other two directions. Similarly, varying the ratio p/q simulates preferred transport toward the absorbing boundary, e.g., along an applied field axis.

Given the lattice anisotropy parameter p + q = 1/3 and the hop bias p/q, the conduction anistropy can be relatively easily evaluated for a system in which carriers are introduced at the origin $X = X_0$, t = 0, and allowed to diffuse toward the absorbing barrier at X = a. Specifically, we can calculate the particle distribution funtion P(X, t) and the distribution of transit times g(t) as a function of lattice anistropy and hop bias. These quantities are given as[3]

$$P(X, t) = \sigma\sqrt{\frac{1}{2\pi t}} \exp - \frac{(x - \rho t)^2}{2\sigma^2 t} - \exp\left[\frac{2}{\sigma} - \frac{(X - 2 - \rho t)^2}{2\sigma^2 t}\right] \quad (8\text{-}5)$$

$$g(t) = \frac{1}{\sigma\sqrt{2\pi t^3}} \exp\left[-\frac{(1 - t)^2}{2\sigma^2 t}\right] \quad (8\text{-}6)$$

where $\rho = p - q$ and $\sigma^2 = p + q - (p - q)^2$

The distribution of charge carriers through the sample at various times for the isotropic case is illustrated in Figure 8-2 This should be compared with Figure 8-3 for the case where the system is still isotropic in hop length, but an applied field makes the probability of a hop right nine times that of a hop left. Although the distribution of particles remains broad, the maximum of that population distribution shifts perceptibly toward the absorbing boundary. However, if the hop length along the x axis is made four times that on the y or z axis (e.g., an assembly of molecules with localized state dimensions 4:1:1), the maximum in P(X, t) shifts dramatically toward the absorbing

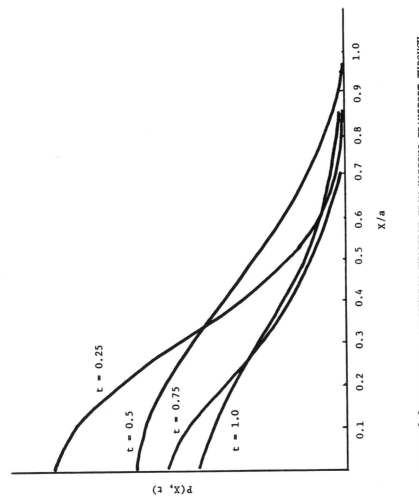

FIGURE 8-2 PARTICLE DISTRIBUTION FUNCTION FROM HOPPING TRANSPORT THROUGH AN UNORIENTED, ISOTROPIC MATRIX

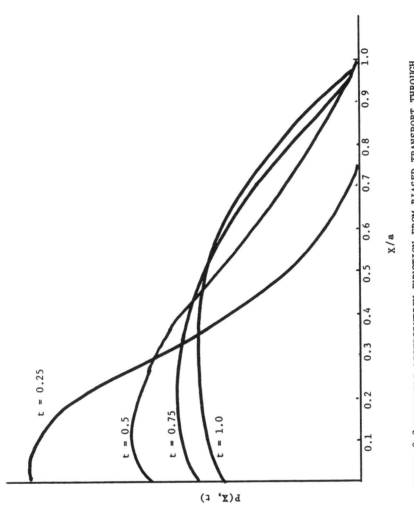

FIGURE 8-3   PARTICLE DISTRIBUTION FUNCTION FROM BIASED TRANSPORT THROUGH
AN ISOTROPIC MATRIX

structure as indicated in Figure 8-4.

These calculations are combined in Figure 8-5 to illustrate the relative effects of orientation and applied field on the transport of charge particles. Plotted are the z axis mobility versus orientation function p + q at various lead bias values p/q. An isotropic specimen is described by a vertical line at p + q = 0.33. Increasing the fraction of hops in the X direction toward the absorbing boundary does, in fact, increase the overall mobility. However, introducing a relatively small degree of anisotropy into the hop length provides a significantly greater increase in mobilty.

Returning to our original question, we ask what transport anisotropies could be expected from relatively small structural differences. For this crude model, the most significant transport anisotropy that can be realized (for a case where all hops take place along the x axis and all go toward the absorbing electrode) is only three. It is informative to compare this number with the conduction anisotropies in Table 8-1 found in the literature. First, values of $\sigma_{11}/\sigma_{\perp}$ of the same order of magnitude are reported for several polymers, including single-crystalline SN_x.[4] Values to orders of magnitude 10-100 are reported for materials such as the polydiacetylenes,[5] square-planar platinum,[6] and TTF-TCNQ[7] mixed-valence molecular crystals. The only really large value reported[8] is probably due to ionic transport and not relevant to this discussion.

What is the molecular origin of these anisotropies? For polyurethanes, TCNQ complexes, and polyacetylene, this anisotropy is derived solely by stretch elongation of amorphous chains and crystalline lamella. In nylon 66, for which conduction is again probably due to ionic transport, the anisotropic structure is due to extensive hydrogen bonding with crystalline regions. In SN_x, however, there is a very high degree of chain order. All chain axes point in the same direction, and extensive registry exists between chains. Yet the relatively low intrinsic anisotropy is indicative of substantial charge transport between chains, which is in contradiction to most models of transport in

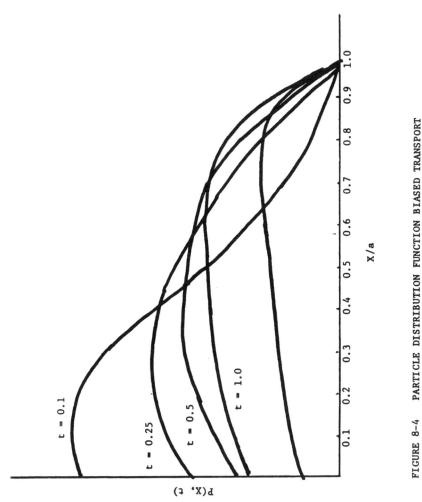

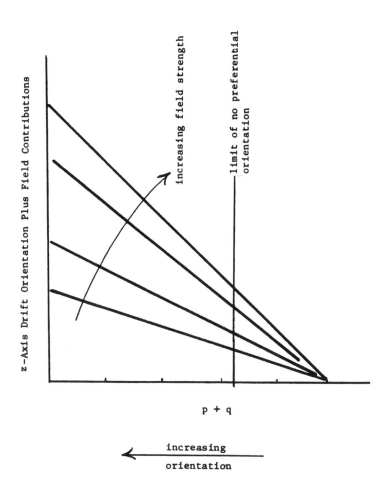

FIGURE 8-5 COMBINED EFFECT OF BIASED TRANSPORT AND LATTICE
ANISOTROPY ON z-AXIS MOBILITY

Table 8-1

CONDUCTIVITY ANISOTROPY IN ORIENTED POLYMERS
AND QUASI-ONE-DIMENSIONAL SOLIDS

Compound	Reference	$\sigma_{11}/\sigma_{\perp}$	Comments
Polyurethane TCNQ complex	9	2.0	Twofold increases in σ at 15% elongation
Nylon 66	10	4.0	Normal to hydrogen bond plane
Polyacetylene	11	10	Undoped, 300% elongation
Mixed-valence square-planar platinum complex	6	200	
TTF:TCNQ	7	400	
Polydiacetylene	5	800 ±300	Photoinduced mobilities
Polysulfur nitride	4	100	300 K
Polybenzobisthiazole	8	∿1,000,000	

quasi-one-dimensional polymers.

The systems with higher anisotropies have as high a degree of order as does SN_x, but the polymer backbone in polydiacetylene is surrounded with a large volume fraction of pendant groups that insulate and separate the delocalized backbones. Similar considerations apply to the transition metal complexes and to the charge transfer salts in which lateral electrostatic interactions between charge transfer sites are repulsive.

The relatively low anisotropies found, even for single-crystalline SN_x, have an important implication for the design of new conducting polymers. Previous experiments have designed molecules based on a model of an elongated, one-dimensional conjugated chain in which electrons traveled freely up and down the backbone. In the most perfect example of this model, SN_x, conduction normal to the axis is only slightly lower than conduction along the axis. Evidently a one-dimensional, delocalized, π backbone is not a prerequisite for high conductivity. Periodic order, however, is required, as evidenced by the stretch-induced anisotropy in semicrystalline polyacetylene. Later in this section, we tie together these considerations to form a comprehensive model for conductivity in organic polymers.

Transport Through Periodic Potential Barriers

We have described the transport of charged carriers by hopping between nearest neighbor sites. Although these simple considerations can explain the relatively low mobilities of molecular crystals, they are inadequate to explain the high mobilities observed in doped polyacetylene, polypyrrole, or polyphenylene sulfide. We have looked for ways to obtain high mobilities from charge hopping in hope of unifying the application of hopping models to low conductivity organic polymers and to reconcile the fact that the temperature dependence of the conductivity in these three highly conducting systems is characteristic of that observed for hopping systems rather than for band gap materials.

A possible mechanism for rationalizing this discrepancy is provided by theoretical calculations of the transmission of electrons through a periodic array of potential barriers. Simply put, for a periodic system of identical barriers, there exists a set of quasi-stationary states or energy levels for which the transmission coefficient through the barrier is unity[12] (this is a quantum mechanical analog to the transmission of light through multilayer interference coatings). More sophisticated calculations[13] have explored the effect of varying the potential well widths, well depths, and barrier widths on transmission coefficients for tunneling. Although more cumbersome experimentally, these calculations reproduce the results of the simpler model in the limit of periodic potential barriers.

We introduce these models to illustrate why we are interested in examining how localized charge states can interact. Assuming that intermolecular localized charge states form around the site of dopant molecules and that these states interact to form long range periodic electrostatic wells and barriers, it is reasonable to expect that macroscopic transport can be characterized by short-range hopping at low dopant concentrations and long-range tunneling, followed by short-range hops between those regions at higher dopant concentrations. Before discussing this model semiquantitatively, we want to introduce the percolation phenomenon as it applies to dopant concentration and the formation of periodic potential barriers.

Percolation Models

The concept of percolation refers to the formation of an infinite network of markers as they are placed onto an empty lattice.[14] The parameters that characterize a percolation network include the volume fraction of occupied sites ϕ, the mean size of clusters of contiguous occupied sites n, and the critical volume fraction ϕ_c, at which a cluster extends from one boundary to another. Values of ϕ_c and the behavior of various properties at volume fractions near ϕ_c are of considerable interest in modeling phase transitions at critical points and represent a body of knowledge that we can use in trying to understand how dopants affect conductivity.

Because this is a lattice model, it may be somewhat artificial, particularly for amorphous systems, because it requires a specific lattice coordination number. The results of percolation models are believed to be nearly independent of the type of lattice used around ϕ_c. The critical volume fraction ϕ_c does vary somewhat according to the type of lattice used. For example, a ϕ_c of 0.198 for a face-centered cubic lattice becomes 0.245 for a body-centered cubic and 0.311 for a simple cubic lattice. In three-dimensional amorphous systems, a value of ϕ_c equal to 0.30 has been calculated.[15] The issue of what is the most appropriate lattice is not relevant to our general discussion, however, and we simply consider ϕ_c to be approximately 0.3.

Because we are also interested in a macroscopic transport process involving long-range hops through a periodic lattice of doped sites, we will also want to calculate the probability of finding a line of occupied sites with a given length x and the mean value of $\bar{x}$. For a simple planar rectangular lattice with a lattice coordination number Z of 4, the probability that any given site is occupied is equal to ϕ. The probability of a contiguous occupied site is then very nearly $Z \cdot \phi$ (or 4ϕ). Thus, because there are 4 possible neighbors and the probability of each being occupied is ϕ, there are now only two possible sites, the probability of each being nearly ϕ. Thus, the probability P_x of finding a linear assembly x units long is very small for $\phi < \phi_c$.

$$P_x = \phi \cdot 4\phi \cdot (2\phi)^{x-2} \quad (Z = 4) \tag{8-7}$$

or generally

$$= Z\phi^2 \cdot (2\phi)^{x-2}$$

The average length $\bar{x}$ could then similarly be calculated as the ratio of the number of linear assemblies n_x of length x divided by the total possible number of assemblies. Estimating the number of possible linear assemblies is cumbersome, but percolation theory tells us that near the percolation threshold, the total number of clusters $\Sigma n_s(\phi)$, and the sum over all cluster sizes S of the product $n_s \cdot s$ are given by the scaling relations

$$\Sigma n_s \sim (\phi - \phi_c)^{2-\alpha} \qquad (8-8)$$

$$\Sigma n_s \cdot S \sim (\phi - \phi_c)^{\beta} \qquad (8-9)$$

where α and β are well-known universal scaling exponents equal to -0.5 and 0.4, respectively. The ratio $\Sigma n_s \cdot S / \Sigma n_s$ is then the number average cluster size $\bar{S}_n$, which allows the estimate of the domain radius $\bar{S}_n$.

$$\bar{S}_n = \frac{\varepsilon n_s \cdot S}{\varepsilon n_s} \quad \approx \quad \frac{(\phi - \phi_c)^{\beta}}{(\phi - \phi_c)} \ (2 - \alpha) = (\phi - \phi_c)^{-2.1} \qquad (8-10)$$

The important feature of this result is that at low occupancies only small clusters and small localized electronic states exist. As the critical volume fraction for the establishment of an infinite network is approached, the size of these states increases exponentially, as shown in Figure 8-6.

Functional Dependence of Conductivity in the Generalized Model

This section brings together our previous observations to form a comprehensive model for conduction in organic polymers. It is comprehensive in the sense that the structural features and behaviors described in the model are applicable to a variety of polymer systems. Key features of the model are summarized as follows:

- Macroscopic charge transport occurs through the hopping of charges between localized electronic states. As the number and volume of localized states increases, the macroscopic conductivity increases rapidly as a continuous percolation network that allows very long-range transport is formed.

- The localized electronic states that serve as lattice points in the percolation network may be solely unimolecular (e.g., TNF-doped polycarbonate), intramolecular backbone segments (e.g., undoped polyvinyl carbazole or pyrolyzed kapton), or

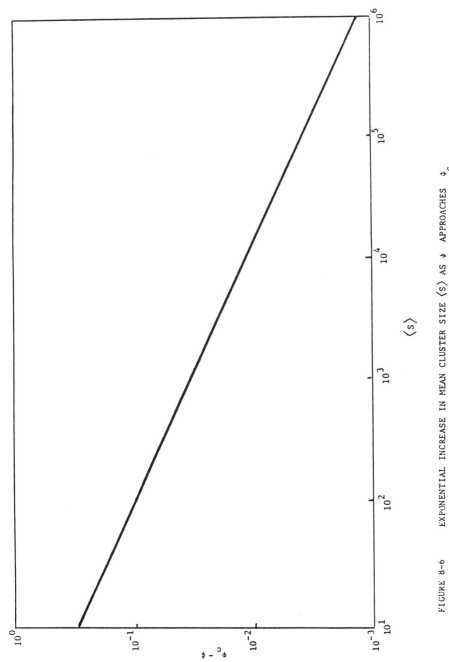

FIGURE 8-6   EXPONENTIAL INCREASE IN MEAN CLUSTER SIZE $\langle s \rangle$ AS ϕ APPROACHES ϕ_c

intermolecular ordered dopant-repeat unit aggregates (e.g., AsF_5-doped polyphenylene sulfide or polyacetylene). The volume and relative electronic energy of these localized states determine the critical number of localized, charged states required for the percolation network to form.

- The hop length depends on the extent of long-range periodic order. In disordered systems with little or no long-range order, hops between adjacent sites are involved. If periodic superlattices are formed, long-range tunneling occurs, which appears as variable-range hopping and leads to the high mobilities necessary for the realization of metallic conductivity.

With this model we can reconcile the diverse phenomena reported in the literature and understand how dopants act in different systems. We also include some model calculations describing the effects of orientation between localized states on conduction anisotropy polymer systems.

Background

In developing this model, we sought to rationalize seemingly contradictory behavior reported in the literature. For example:

- The temperature dependence of the conductivity for all the materials previously discussed, excluding those that are ionic, excitonic at very high fields, and metallic (SN_x), was thermally activated in accord with either an adjacent site or variable-range hopping mechanism or tunneling between metallic grains. However, the thermopower, which does not involve charge transport, is typical of band gap conduction in metals.

- None of the polymers studied is of a high enough molecular weight for a single delocalized backbone to be responsible for macroscopic charge conduction, requiring that intrachain transport must be operative and be the rate-limiting step. Yet popular theories relate bulk conductivity to calculations based on the molecular structure of a single repeat unit.

- Doping of polyacetylene, polypyrrole, and polyphenylene sulfide and cofacial polyphthalocyanines with strong oxidizing agents at a mole ratio level of 1:10, 1:4, 1:1, respectively, brought about an immediate increase in conductivity of 10 to 13 orders of magnitude. Similar effects, at 1:1 ratios, are observed for charge transfer or molecularly doped complexes with iodine,

TCNQ, trinitrofluorenone, and triphenylamine.

- Significant increases in conductivity are observed on the formation of very large fused-ring systems from pyrolyzed Kapton, polyacrylonitrile, and polyacenequinone polymers. Doping is neither necessary nor sufficient for high levels of conductivity to be realized.

Although specific theories exist for each of these materials independently, they are of little use in identifying new materials or interpreting those that do not fit into a previous mold.

The Model

The basic structural element of our model is the localized, inter- or intramolecular charge-stabilizing site. Because this can take on so many different forms, we treat it simply as a potential well. That is, it may be a single molecular dopant in an inert matrix or a three-dimensional assembly of an inorganic dopant with unsaturated repeat units from the backbone of several different polymer molecules. This structural element can be characterized by three specific properties: the volume v, the well depth (barrier height) V_o, and the energy E of the highest energy electron in the well corresponding to the energy of the highest occupied molecular orbital in molecular orbital theory.

Determining the volume of the localized state is difficult except when that state is well defined, e.g., a bimolecular charge transfer complex or a molecular dopant dispersed in an inert matrix. We can arrive at a functional description, however, using the criterion that the onset of high conductivity in doped polyacetylene, polypyrrole, or polyphenylene sulfide occurs when the total volume fraction of polymer-dopant aggregates exceeds the percolation threshold. Thus, the localized state volume is given by

$$v = \phi_c = \frac{\text{Volume of the system}}{\text{Number of dopant molecules}} \qquad (8\text{-}11)$$

or in terms of the number of polymer repeat units to dopant molecules

$$v = \phi_c \quad \text{(molar volume/repeat unit)} \times \text{(repeat unit/dopant molecule)}$$

Both experimental and theoretical considerations can be applied to evaluate the energy required to remove an electron from the localized state. The most readily applied method is to calculate an activation energy based on the temperature dependence of the conductivity. A second experimental method is to use photoelectron spectroscopy to measure the energy required to excite electrons into a continuum. Theoretically, one can use particle-in-a-box methods to calculate the relative energies of n quantized orbitals and, by filling each orbital with two electrons per bond, estimate the energy E of the highest occupied molecular orbital. Frequently, the barrier height V_o is assumed to be twice this value.

The Functional Dependence of the Conductivity

With the mobility concepts and definitions of molecular structural properties, we can then calculate the conductivity--or at least the dependence of the conductivity on structural and environmental variables. For this we need to know the number of carriers, their charge q, and their mobility μ. For electronic carriers, q is equal to $|e|$, the charge of an electron.

Similarly, the number of carriers is equal to the number of localized electronic states formed. For doped polymers (I_3^-, $A_sF_6^-$, BF_4^-), this number is proportional to the dopant concentration. In charge transfer salts there is one localized state per charge transfer pair. Similarly, for partially oxidized cofacial polyphthalocyanines, there is a localized state for every metal atom-phthalocyanine dopant assembly. In pyrolyzed Kapton or polyacylonitrile, counting the number of states is more difficult because they are ill-defined π orbital "condensates." They could be estimated by counting the number of free electron spins evidenced in the ESR spectrum, but on pyrolysis the molecular structure, density, resistivity, and weight all change, making correlations difficult.

Calculating the mobility is the critical issue. With our approach we define two mobility states, one above and one below the percolation threshold ϕ_c. Below ϕ_c, μ is given by equation (8-3) as μ = e D/kT for nearest neighbor hopping and D is Nl^2. The hop length l is given by geometric considerations about the average distance between localized states or the average cluster radius from the percolation approach. Calculating the number of hops per unit time 1/N then depends on the transmission coefficient for an electron to leave the localized state and the frequency ν with which an electron tries to escape the potential well.

Transmission out of a potential well was given in equation (8-12) as a function of the well properties.

$$T = \exp -2\sqrt{2m(V_0 - E)/h^2} \ \cos^{-1}\left[\sqrt{a/b} - \sqrt{a^2/b^2}\right] \qquad (8-12)$$

As we noted in a previous section, the probability of transmission reaches 87% of its theoretical maximum when the carrier width equals the well width, and increasing it beyond that point leads to only a marginal increase. The frequency with which an electron hits the wall, for systems in which the mean free path is comparable to the localized state dimension, is inversely proportional to the distance that the electron has to travel before it encounters a barrier and so varies as $v^{-1/3} \sim s$. This is a rather weak dependence and is extremely sensitive to the local atomic structure. Because the localized states in most currently known materials are roughly the same size, we set ν equal to unity and accept that the probability of a hop is roughly equal to the probability of transmission out of the well.

Thus, well below ϕ_c the most important variable on which conductivity depends is the energy difference between the barrier height and the particle energy. The functional form of the conductivity is then

$$\sigma \sim \phi_d \cdot \exp(1/kT) \qquad (8-13)$$

$$\sigma \sim \phi_d \cdot \exp(V_o - E/kT) \qquad (8\text{-}13)$$

At these low dopant levels, only very small clusters are found. As ϕ_c is approached, the size of the localized status (equivalent to the hop length) increases exponentially, whereas the frequency with which an electron hits the wall and the probability of it escaping vary only slowly. In addition, the well width is increased, decreasing the $V_o - E$ difference and raising T. The dependence of the mobility on dopant concentration is then governed by the exponentially increasing hop length. Phenomonologically, this introduces a variable-range hopping mechanism and the concomitant fractional temperature exponent. The functional form of σ then becomes

$$\sigma \sim \exp\left[(V_o - E/kT)^y\right]\left[(\phi - \phi_c)^{-6.3}\right] \qquad (8\text{-}14)$$

which is dominated by the $(\phi - \phi_c)$ term and the fractional temperature exponent.

Above ϕ_c there are large extended delocalized regions through which electrons travel freely. Structural order is not really long range, however, and macroscopic transport is limited by a thermally activated variable-range hopping, although mobilities are significantly increased by the extended, yet localized, charge states.

$$\sigma \sim \exp\left[(\text{const}/T)^{1/y}\right], \qquad y = 2 \text{ to } 4 \text{ depending on } \phi > \phi_c \qquad (8\text{-}15)$$

Discussion

This model has specific limitations. It is phenomonological in nature and is not an **ab initio** quantum mechanical calculation of transport in disordered systems. It does, however, reorganize intermediate degrees of order that are important features of polymer systems and points out the molecular nature of those active species responsible for conduction, a severe task to complete rigorously. It also succeeds in phenomonologically modeling conduction in polymers in a

general sense and rationalizes the diverse temperature and dopant
dependencies of the conductivity. Finally, it has succeeded where no
others have in identifying new polymers with unusually high
conductivities.

References

1.   M. Karplus and R. Porter, Atoms and Molecules (W. A. Benjamin, Menlo Park, CA, 1970).

2.   H. Pohl, A. Rembaum, and J. Moacanin, "Polymeric Semiconductors," in Progress in Dielectrics, Vol. 6 (Academic Press, New York, 1965).

3.   D. Cox and H. Miller, Theory of Stochastic Process (J. Wiley, New York, 1965), p. 22.

4.   P. Barrett et al., J. Phys. F11 (3), 557 (1981).

5.   K. Lochner et al., Chem. Phys. Lett. 41 (2), 388 (1976).

6.   M. Rice and J. Bernasconi, J. Phys. F2 (5), 905 (1972).

7.   L. Brickford and K. Kanazawa, J. Phys. Chem. Solids 37, (9) 839 (1976).

8.   R. Barker and D. Chen, Proceedings of the Conference on Electrical Insulation and Dielectric Phenomena (National Research Council, Bostom, MA, 1981), p. 297.

9.   R. Landel et al., J. Poly. Sci. Lett. B9 (8), 627 (1971).

10. D. Seanor, J. Poly. Sci. Symposia 17, 195 (1967).

11. Y. Park et al., J. Chem. Phys. 73 (2), 946 (1980).

12. E. A. Pshenichonor, Soviet Physics Solid State 4 (5), 819 (1962).

13. F. L. Carter, "Electron Tunneling in Short Periodic Arrays," in Molecular Electronic Devices, F. L. Carter, Ed. (M. Deckker, New York, 1982).

14. D. Stauffer, Physics Reports 54 (1), 1 (1979).

15. N. Economu, M. Cohen, K. Freed, E. Kirkpatrick, "Electronic Structure of Disordered Materials," in Amorphous and Liquid Semiconductors (Plenum Press, New York, 1974), p. 101.

9. Design and Preparation of New and Modified Electrically Conducting Polymers

The relationships between molecular structure, mechanical strength and environmental stability are well known. In contrast, as the material in the previous sections indicates, the relationship between molecular structure and electrical conductivity is complex. Thus, in principle, there should be an intelligent method by which to modify materials to improve performance or to prepare new materials with the requisite combination of mechanical, thermal, environmental, and electrical properties required for spacecraft use.

Molecular Features that Enhance the Quantity and Mobility of Charge Carriers

With the model of conductivity presented in Section 8, we can identify structural features that should contribute to conductivity. In principle, materials with variable-range hopping, band-gap semiconduction, and metallic conductivity can be designed.

Variable-range hopping is associated with truly amorphous structures and low mobilities. The charge carriers are extremely localized and so must be of relatively high energy and stability. Little distinction need be made with respect to residues covalently bound to the polymer as pendant groups or within the backbone. Many such stable radical anions or cations are known in the literature and could be relatively easily incorporated into a variety of polymer structures.[1] Perhaps the most important criteria in this regard is that the matrix be chemically inert to the reactive, radical species.

Mixtures of molecules that can be photoexcited into radical ions have been widely investigated.[2,3] The relatively high fractions of low molecular weight material required have led to serious degradation of the matrix mechanical properties. Work with polymer bound charge carriers has focused on the use of TTF or TCNQ, which together form

highly conducting molecular crystals.[4,5] These efforts are generally
not considered successful because they did not create highly conducting
materials, but it is perhaps unreasonable to have expected it.   This
approach is fairly well proven for producing low-conductivity, low-
mobility materials with readily variable mechanical properties. New
directions in the field include the use of other well-known, (outside of
conducting polymers) stable, radical ion groups and further tailoring of
the polymer thermal and mechanical properties to specific applications.

A clear distinction may be drawn between the design of materials
with hopping and band-gap conduction mechanisms. A less clear
distinction may be drawn between band-gap materials with metallic and
band-gap materials with semiconducting properties. The two independent
parameters available are the extent of superlattice formation and the
effective size of the lattice element. For example, if a fully
developed superlattice is present, then the conductivity will depend on
the relative energy and hence the effective size of the localized
state. Alternatively, for a group of localized states with the same
energy, their number and the extent to which the lattice is occupied
will determine the conductivity.

Let us first examine the factors, both intra- and intermolecular,
that define the localized state. Electron delocalization relies on the
orbital overlap of periodic distribution of atomic orbitals. The
intramolecular structure should therefore be stiff locally, encouraging
electron delocalization within the backbone, in the traditional sense.
Extensive delocalization along a chain however leads to severe
processing problems.

Of equal importance, although uniformly ignored in the literature,
is intermolecular order. All the polymers that achieved fairly high
levels of conductivity are fairly crystalline. This does not mean that
single crystals, or even a high crystallinity, is necessary, but there
must be sufficient local order present for electron delocalization to
occur, both transverse and parallel to the chain axis. Highly regular
structures are desirable to achieve this order; pendant groups, unless

strongly interacting intermolecularly, should be avoided. Note that we are not requiring long-range delocalized molecular orbitals, only sufficient order for spatially localized, delocalized electron states to form between repeat units on adjacent chains.

The final variable is the electron population of individual localized states and their relative energies. This determines whether conduction will occur through a percolation network, thermally activated band-gap semiconductor, or a metal-like conduction mechanism. The electron population will depend largely on dopants, which create localized charge states by selectively reducing or oxidizing the locally ordered structures. Several dopants are widely reported in the literature. Although they are relatively corrosive, they form stable adducts and hence need not be discounted. Further work would be fruitful in the preparation of intramolecularly doped polymers for which no further dopant is needed.

Perhaps the most difficult concept of this model to engineer is the relative energy, and hence the interaction length, of the localized states. Considerable effort has been devoted to calculating one electron energies for a small section of a polymer or an ideal delocalized segment, but little or no work has been done, to our knowledge, on the interaction of close-packed nearest neighbors on electronic states. Several authors have commented on the importance of this interaction on one electron orbitals,[6] but it is difficult to deal with theoretically. One useful experimental tool that can begin to characterize these states is photoelectron spectroscopy.[7] However, to date it has been used primary with molecularly doped systems rather than on those with the order and electron delocalization we believe necessary for high levels of conductivity.

New Polymers for Spacecraft Applications

Some structural modifications that can be made to improve the thermal and mechanical properties of these polymers with conductivities in excess of 10^{-12} (ohm cm)$^{-1}$ are shown in Figure 9-1. For example, the

I

R_1 = $-CH_3$
$-\emptyset$
$-H$

II

R_2 = $-\bigcirc- C\equiv CH$

$-\bigcirc- CH_2OC-CH=CH_2$
$\quad\quad\quad\quad\quad\overset{\|}{O}$

$-C\equiv CH$

$-C-CH_2-C\equiv CH$

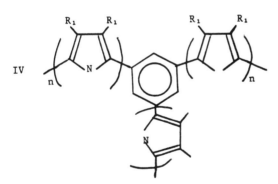

III

n = 4 to 10

IV

FIGURE 9-1   STRUCTURAL MODIFICATION OF THE POLYPYRROLE BACKBONE

substituted pyrroles (I) have the necessary conductivity but lack strength and thermal stability. Adding reactive pendant groups (II) to a homopolymer or copolymer with I would not interfere with the polypyrrole synthesis, but would allow intramolecular crosslinking at elevated temperatures. Replacing R_1 = H with $-CH_3$ or $-\emptyset$ reduces the local order and lowers the conductivity so a similar effect will probably be observed for each R_2 suggested.

Since the substituted pyrroles are known to readily copolymerize, one can envision an acetylene-terminated polypyrrole (III) that, for small n, would probably be soluble in organic solvents. Heat treatment of III will lead to addition across the terminal $-C\equiv CH$ groups, leading to a strong, insoluble, thermally stable matrix (IV) or analogous structures. The networks prepared from these low molecular weight acetylene-terminated resins are becoming very important commercially and are of considerable interest to the aerospace industry.

The pyropolymers require a somewhat different structural modification if they are to be useful in moderating the spacecraft-charging problems. Partial pyrolysis of Kapton may lead to the desired conductivity range, but it is uncertain whether or not these materials would retain sufficient mechanical properties or whether or not they would continue to pyrolyze in space. One possible solution to these problems is to copolymerize the pyrolyzable aromatic polyimide backbone (Va) with a more thermally stable heterocyclic polymer (Vb) as indicated in Figure 9-2. Depending on how the polymerization is achieved, n and m may range from 3 or 4 to 10 or 50, meaning that some segregation of repeat unit types would occur and that there would be enough Kapton-type material to form the large fused-ring systems responsible for conductivity in pyrolyzed Kapton.

An alternative to copolymerization would be the fabrication of polymer blends or alloys. The use of polymers such as Vb as thermally stable reinforcement for thermoplastics, or other less stable polymers, has been discussed for several years. Ideally, the pyrolyzed Va would then be mechanically reinforced by a web-like network of Vb.

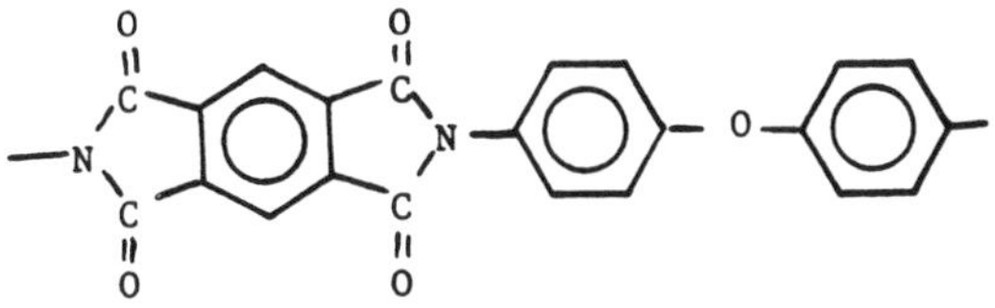

+ Vb

Va

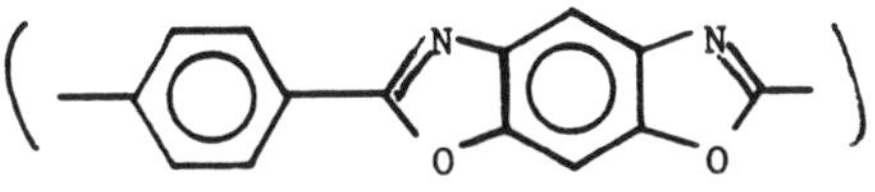

Vb

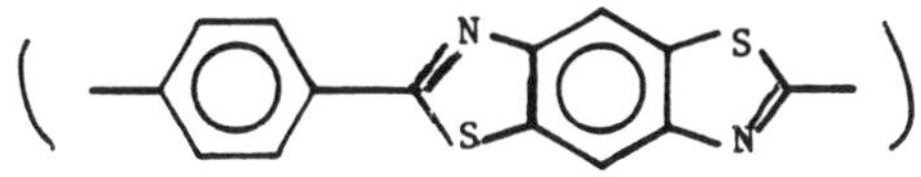

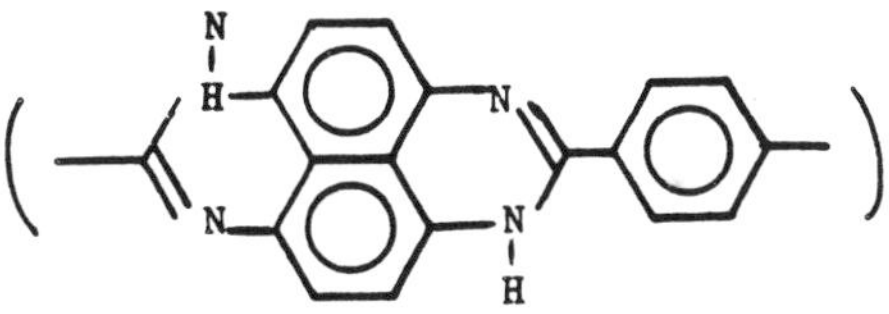

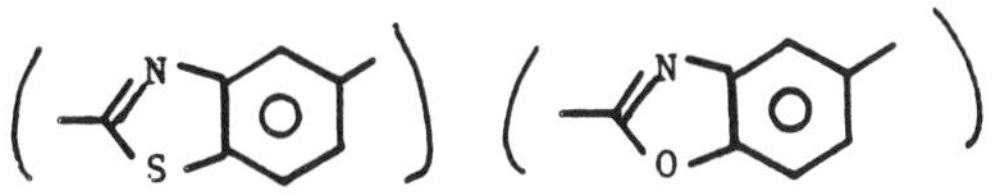

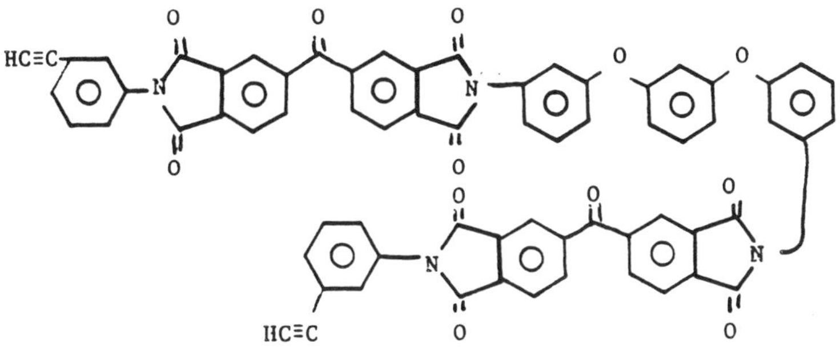

VI

FIGURE 9-2 THERMALLY STABLE HETEROCYCLIC SEMICONDUCTING POLYMERS

A similar approach would be to use the commercially available acetylene-terminated aromatic polyimide resin VI (Thermid 600). This material is used primarily as strong, tough, high temperature coatings, and it may be possible to prepare films from this resin with mechanical properties comparable to those of Kapton and pyrolyze them to obtain conductive materials.

The other well-known material that may be useful in spacecraft charging is polyvinyl carbazole (VII) (see Figure 9-3). The commercially available material appears to meet the mechanical and conductivity requirements, and it is almost suitable for 250°C service. One useful modification would be to identify a catalyst system capable of preparing isotactic PVK, which would be more highly crystalline, have a higher melting point, and be stronger than the commercially available material. An alternative route would be to prepare crosslinkable derivatives of VII though either copolymerization (VIII) or substitution with a reactive group (IX). Of course, this will reduce the crystallinity, but highly crosslinked systems will still possess high thermal and mechanical stability.

It may also be possible to prepare ionomer analogs of PVK. A few percent of ionic, thermally reversible, crosslinks can increase the glass transition temperature of ionomers by 50° to 100°C. Thus, a few percent of a carboxylic acid or sulfonic acid containing pendant groups copolymerized with PVK (X) and complexed with a divalent cation, e.g., Cu^{2+} or Zn^{2+}, could provide the extreme thermal stability necessary for 250°C service.

Further attempts could also be made to produce nonvolatile charge-transfer complexes. If both donor and acceptor species were polymer bound, as illustrated in Figure 9-4, then the probability of stability toward high vacuum and possibly toward elevated temperature would be enhanced. The reactions indicated in Figure 9-4, would allow considerable latitude in modifying the polymer backbone and in preparing crosslinked materials to achieve the requisite mechanical properties.

Our efforts to understand the phenomenon of conduction have also

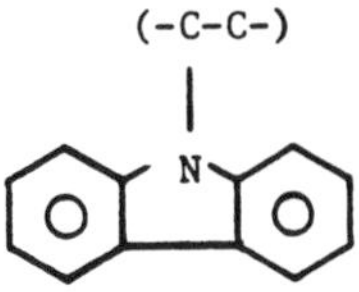

VII

VIII

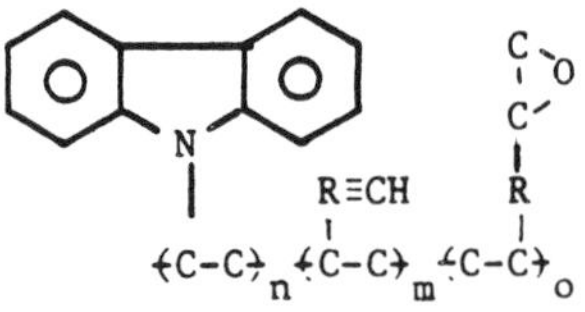

IX

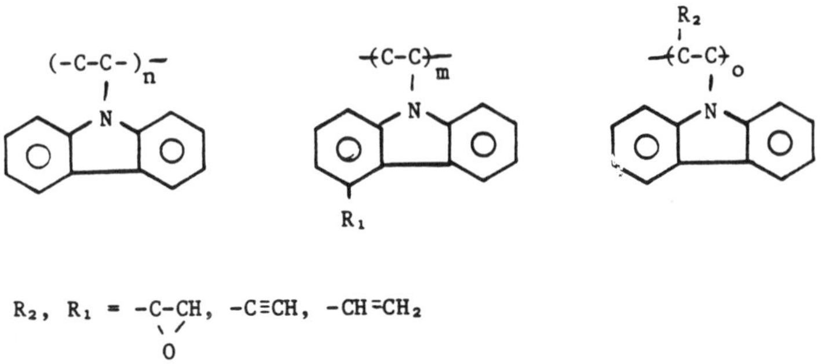

X

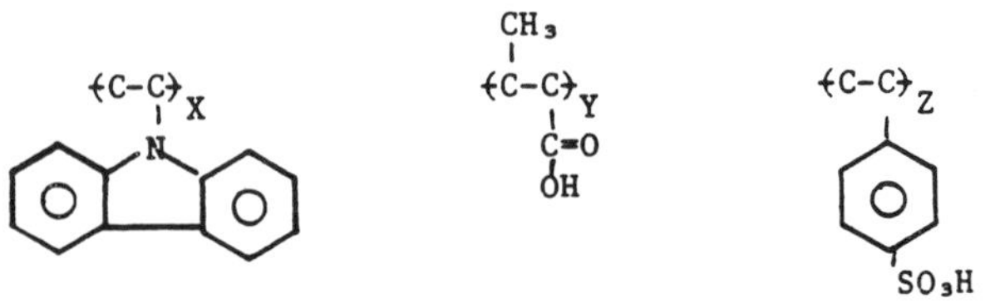

X>>Y,Z

FIGURE 9-3 STRUCTURAL MODIFICATION OF POLYVINYL CARBAZOLE

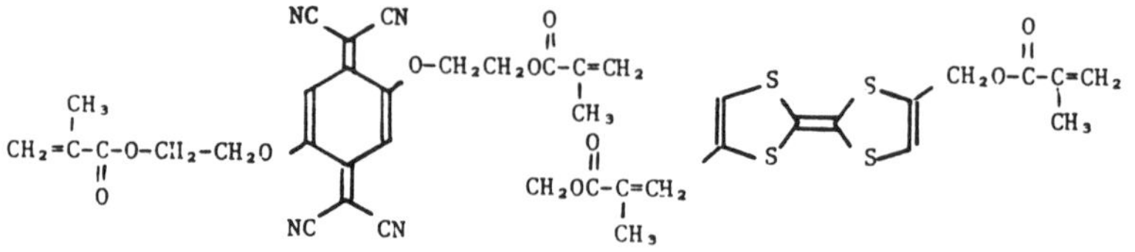

Copolymerization or Homopolymerization Followed by Blending

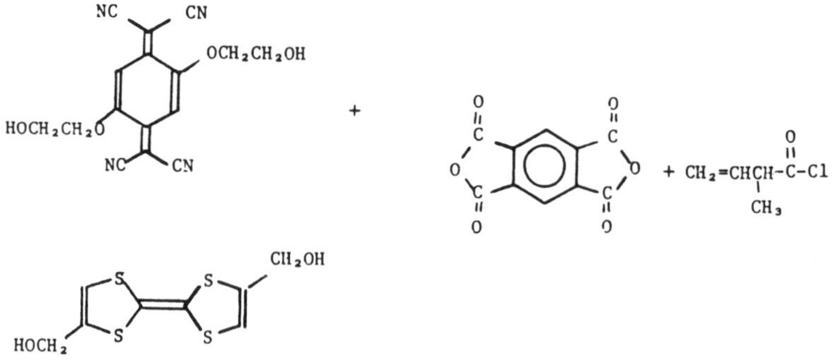

FIGURE 9-4   CHARGE TRANSFER POLYMER BLENDS

led to the identification of several new materials that we believe have
the requisite periodic structural order and stability necessary for the
development of high levels of conductivity. The structures we have
identified are based on well-known highly colored dyes and stable
radical ions that can be incorporated into stereoregular polymer
backbones with few or no pendant groups. Some of the repeat units and
the stable ions that can be prepared by reduction or oxidation of those
repeat units are shown in Figure 9-5.

The radical species associated with each of these violene and
azoviolene structures obtained by the two-step oxidation of the neutral
compound

$$X(CH=CH)_nX \longleftrightarrow X(CH=CH)\overset{\bullet}{X}^{\oplus} \longleftrightarrow \overset{\bullet}{\oplus}X(CH=CH)X^{\oplus}$$

can be characterized by having an odd number of π electrons distributed
over an even number of main chain atoms. Although long-range
delocalization is unlikely, it would be analogous to a conduction hole
in the valence band.

A completely analogous situation exists for systems containing an
even number of π electrons on a odd-numbered backbone, e.g., an unpaired
electron in the conduction band. Some possible structures of this type
$X(CH=CH)C=\overset{\bullet}{X}^{\ominus}$ are summarized in Figure 9-6.

Similar opportunities exist for using the unsaturated heterocyclic
backbone of the aromatic carbodiimides as shown in Figure 9-7. Although
less is known about the conductivity or electrochemistry of these
polymers, they are recognized as being highly crystalline, mechanically
strong, and thermally stable.

It may not be possible to prepare polymers with all these
structural units, but model compounds for all of them are known. Their
redox chemistry has been fairly well studied, which leads to one final
observation: the same redox chemistry that makes electrochemical cells
using polyacetylene possible can be realized with a multitude of other
polymer chemistries. Furthermore, because a range of redox potentials

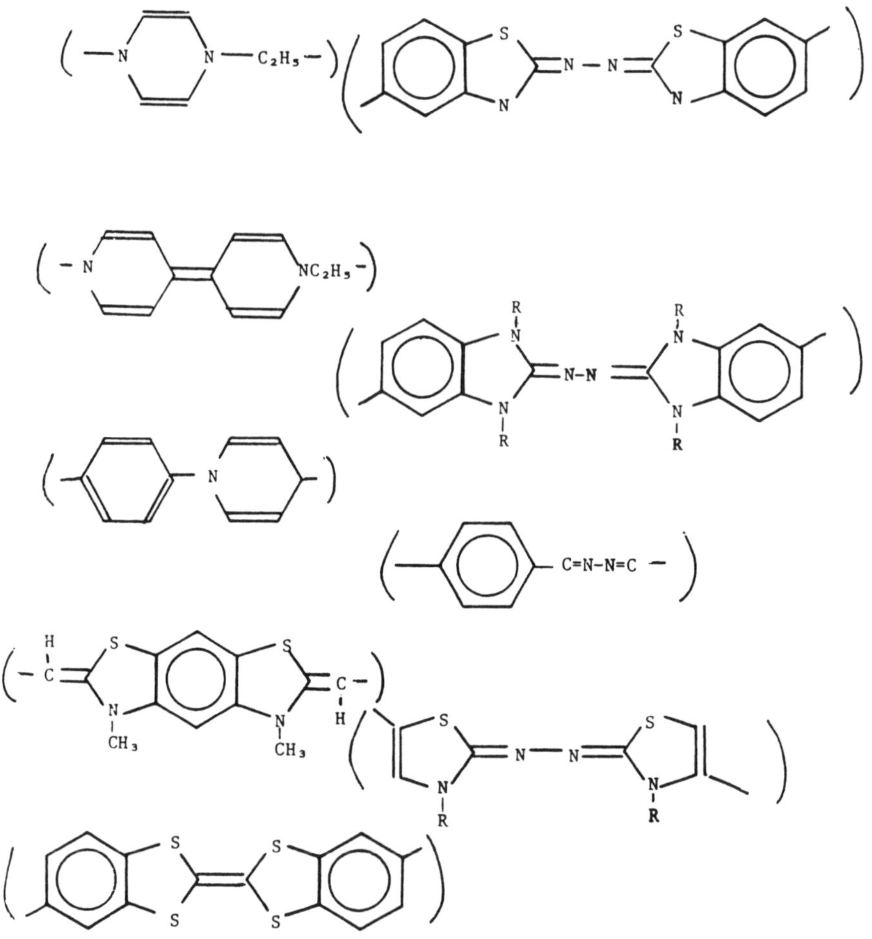

FIGURE 9-5   STRUCTURES OF VIOLENE AND AZOVIOLENE POLYMERS CONTAINING
STABLE RADICAL IONS

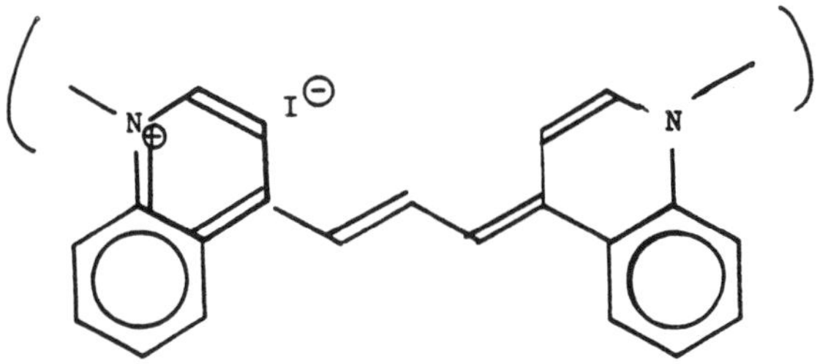

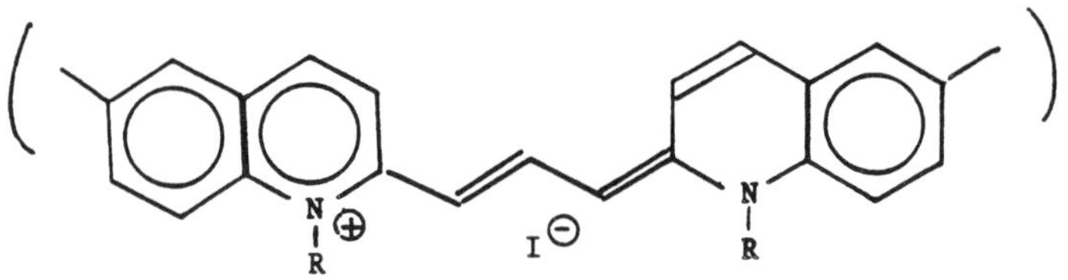

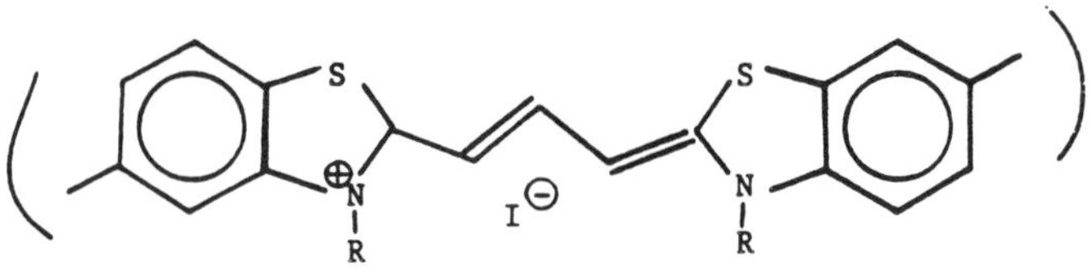

FIGURE 9-6 POLYMER STRUCTURES CONTAINING CYANINE DYE MOIETIES

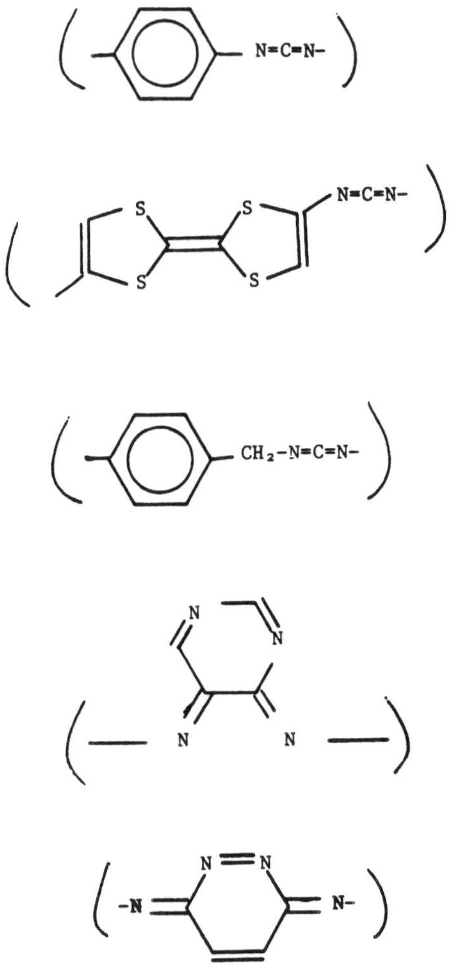

FIGURE 9-7   HETEROCYCLIC AROMATIC STRUCTURES CAPABLE OF FORMING
            LOCALIZED ELECTRONIC STATES AND ORDERED CLOSE-PACKED
            STRUCTURES

are involved, systems can be prepared that are stable with regard to air and moisture exposure. This observation remains to be demonstrated, but the realization of environmentally stable polymeric conductors and battery systems could have a significant impact on aerospace systems.

Synthesis of New Electroactive Polymer Systems

In the previous section, we identified several new or modified polymer structures that we believed would have either an unusually high level of conductivity or useful combinations of thermal, mechanical, and electrical properties. We prepared three of the polyazines and characterized their conductivities, molecular characteristics, and mechanical properties.

The common structural feature of the polyazines is the presence of the chemical structure ($=N-N=$) in the polymer backbone. In general, these polymers are prepared by the condensation of hydrazine (H_2NNH_2) with dialdehydes XI or diketones XII leading to structures XIII and XIV, respectively.[8]

$$
\begin{array}{cc}
\underset{\displaystyle \text{XI}}{\overset{\displaystyle \underset{\text{H}}{\overset{\text{O} \quad \quad \text{O}}{\diagdown \; \diagup}} \; \text{CRC} \; \diagup \diagdown}{}}
&
\underset{\displaystyle \text{XII}}{\overset{\displaystyle \underset{\text{R'} \quad \quad \text{R'}}{\overset{\text{O} \quad \quad \text{O}}{\diagdown \; \diagup}} \; \text{C-R-C} \; \diagup \diagdown}{}}
\end{array}
$$

$$
\begin{array}{cc}
\underset{\text{XIII}}{\overset{\text{H}}{(-C=N-N=C-R-)} + H_2O}
&
\underset{\text{XIV}}{\overset{\text{R'} \quad \text{R'}}{(-C=N-N=C-R-)} + H_2O}
\end{array}
$$

The specific dialdehyde compounds used are glyoxal XV, terephthaldehyde XVI and gluteraldehyde XVII leading to polymers XVIII, XIX and XX respectively.

$$
\begin{array}{cc}
\underset{\text{XV}}{n \;\; \overset{\text{O O}}{\overset{\| \; \|}{H-C-C-H}} + n \;\; H_2NNH_2}
&
\underset{\text{XVIII}}{\overset{\text{H} \quad \text{H}}{(-C=N-N=C-)_n} + 2n \; H_2O}
\end{array}
$$

$$n \ \underset{O}{\overset{O}{HC}}-\hspace{-2pt}\underset{}{\bigcirc}\hspace{-2pt}-\overset{O}{C}-H \ + \ n \ H_2NNH_2 \ \longrightarrow \ XIX$$

$$(-\hspace{-2pt}\underset{}{\bigcirc}\hspace{-2pt}-C=N-N=\overset{H}{C}\!)_n \ + \ 2n \ H_2O$$

$$n \ HC-\underset{H}{\overset{O \ \ O}{\overset{||}{C}}-CH} \ + \ n \ H_2NNH_2 \ \longrightarrow \ (=C-\underset{H}{\overset{H}{C}}-C=N-N=)$$

XVII   XX

To our knowledge, XVIII and XX have not been reported in the literature, although XIX and similar aromatic (benzene-ring-containing) polymers have been investigated[9] and polymeric aducts of formaldehyde with hydrazine have been claimed.[10] We have studied the polymerization of XVIII in several solvents including dimethyl formamide (DMF), dimethyl acetamide (DMAc), benzanilide (BA), water, and dimethyl sulfoxide (DMSO). In DMF and water the reaction is extremely fast, with hydrazine being consumed as fast as it is added to the reaction vessel. In these solvents, however, the polymer precipitates from solution at about the trimer to pentamer (n = 3 to 5) stage, limiting the molecular weight that can be prepared. An IR spectrum of XVIII prepared in this way (Figure 9-8) clearly shows the presence of both unreacted aldehyde and amine and groups as indicated. In DMAc the reaction proceeds more slowly, requiring approximately 24 hours to progress to the point where a methanol-insoluble (but still DMAc-soluble) polymer is formed. In contrast, the reaction in DMSO proceeds to high molecular weight, without loss of solubility, in approximately 1 hour.

The ability to prepare high molecular weight, soluble, semiconducting or conducting polymers would be a significant advance in

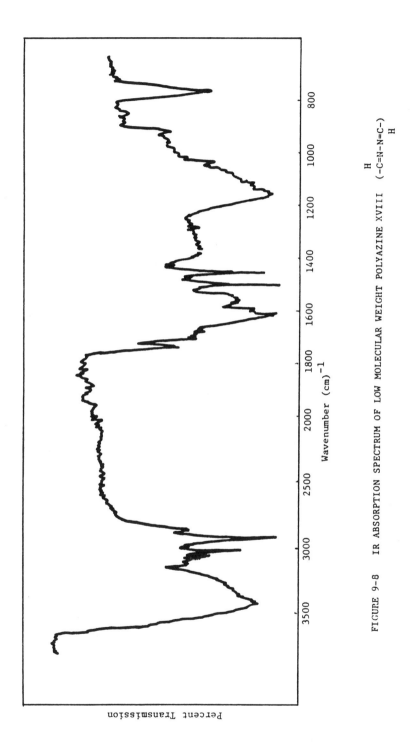

FIGURE 9-8   IR ABSORPTION SPECTRUM OF LOW MOLECULAR WEIGHT POLYAZINE XVIII   (−C=N−N=C−)

this field because the absence of solubility precludes molecular characterization and severely complicates processing.

Preliminary electrical conductivity measurements have been performed using a two-contact, pressed-powder method. The pristine polymer is nonconducting with a resistivity in excess of 10^8 ohm-cm, the sensitivity limit of our instrument. The polymer readily absorbs iodine, however, reaching saturation at about one dopant molecule (I_3-) per ten repeat units. The addition of iodine lowers the resistivity at least four orders of magnitude to a value of 2.35×10^4 ohm-cm.

The effect of iodine addition is further illustrated in the IR absorption spectra (Figure 9-9) of the doped polymer. The characteristic, sharp -C≡N- absorptions at 1200 cm^{-1} are broadened, indicating that on doping the backbone has been partially and locally oxidized and that a distribution of bond strengths is now found instead of a single bond strength. Similar changes in the IR absorption spectra of polyacetylene on doping have been reported.[11]

Solution characterization XVIII has yielded some unusual information. Because this polymer is prepared by a polycondensation reaction, the molecular weight is expected to remain below 50,000. Size exclusion chromatography analysis in DMSO, however, indicates a weight average molecular weight of 510,000, based on polystyrene standards. Two possible explanations for this discrepancy can be imagined. The first presumes that the polymer structure can be modeled as a thin rigid rod as opposed to a random coil-like polystyrene. A rigid rod of molecular weight 1,000 could in fact separate on the column at the same point as a polystyrene of 100,000, but the 120° C=N-N bond angle makes this interpretation uncertain. The second interpretation postulates a strong inter- and intramolecular electrostatic interaction between repeat units. Although we have observed this type of interaction in DMSO solutions of charged polyelectrolyte molecules, it would require the polymer to spontaneously ionize, which is unexpected.

One approach to rationalizing this discrepancy is to combine measurements of the molecular weight with the intrinsic viscosity [n],

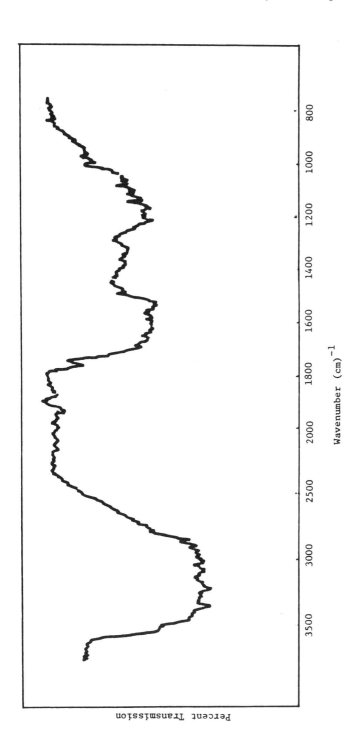

FIGURE 9-9   IR ABSORPTION SPECTRUM OF POLYMER XVIII DOPED WITH IODINE, SHOWING
PARTIAL OXIDATION OF THE C≡N BOND

which is sensitive to both molecular weight, chain conformation, and electrolyte effects. In DMSO the intrinsic viscosity for this polymer is 1.2 dL/g, a fairly high value typical of extended chain conformations, and no polyelectrolyte effects are observed.

If in fact the polymer were rodlike, we can obtain a molecular weight from the intrinsic viscosity based on a theoretical model for length-to-diameter ratios L/d > 50. The intrinsic viscosity can be approximated as[12]

$$[\eta] = 5.81 \times 10^{20} \; \frac{(d_H)^{0.2}}{M_L} \; L^{1.8} \tag{9-1}$$

where d_H, M_L, and L are the hydrodynamic radius, mass per unit length, and length, respectively. Entering reasonable values for these variables in equation (9-1), we arrive at a weight average molecular weight of 9,720, corresponding to a conversion of 98.8%, which is very reasonable for condensation polymers.

Similar polymerizations have been performed with terephthaldehyde and hydrazine. In each of the solvents we studied, the polymer precipitated from solution before high conversions could be reached. The infrared absorption spectra of XIX is shown in Figure 9-10. Similar changes in the shape of the C=N absorption band at 1200 cm^{-1} on exposure to iodine are observed. Pressed-powder conductivity measurements on iodine-doped samples yield conductivities in the range of 10^{-10} to 10^{-8} (ohm cm)$^{-1}$. Because this material is insoluble and does not melt, it would be extremely difficult to obtain molecular weight data. Based on the limited solubility in the polymerization media and the carbonyl endgroups in the IR spectrum, a value of 1000 would not be unreasonable to assume.

The polymer prepared from gluteraldehyde XX has some equally unusual properties. With a partially saturated backbone, it does not have the delocalized π electron backbone of XVIII or XIX. It can also be polymerized to moderate (25-50,000 daltons) molecular weight but is

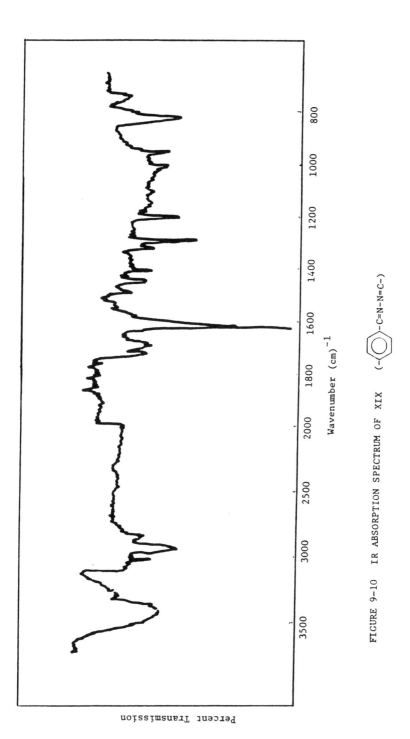

Percent Transmission

FIGURE 9-10   IR ABSORPTION SPECTRUM OF XIX   (-⟨◯⟩-C=N-N=C-)

coil-like in solution compared with XVIII. With the five carbon
backbone repeat units, its morphology is much like that of a highly
crystalline polyethylene. Like polyethylene, however, its conductivity
is low, lying below 10^{-10} (ohm cm)$^{-1}$, the maximum sensitivity of our
apparatus.

The general synthesis method used for each of these three
compositions has been used to prepare copolymers with differing degrees
of backbone stiffness, π electron delocalization, and conductivity.
Copolymers with varying mole ratios of XV, XVI, and XVII have been
prepared in DMSO. Extensive characterization has not been performed,
but moderately high molecular weights were obtained, indicating that the
different monomers will react together in high yield.

A comparison of the molecular structure and properties of the three
homopolymers, and their copolymers, prepared in this study illustrates
the importance of the three structural features, discussed in Section 8,
that we believe are important in preparing soluble, processable, stable,
electrically conducting polymers. The relevant properties for the
compounds we have made are summarized in Table 9-1 along with our
criteria for conductivity: stable localized molecular orbitals capable
of forming radical ions, local chain stiffness but sufficient
flexibility for solubility, good intermolecular packing density.

All of the polyazines are inherently stable and can form an air
stable radical ion when treated with a suitable oxidizing agent, which
is not true of materials like polyacetylene. The reason for this lies
in the difference between the redox potential of the azine group (<1.5V)
and that of polyacetylene (>3.0V).

None of the three homopolymers have bulky pendant groups to inhibit
close packing and intermolecular orbital overlap. Because they are
chemically inhomogeneous the three copolymers are of lower density and
lower crystallinity. Molecular models indicate that the relative order
of backbone flexibility of the three homopolymers is XX > XVIII > XIX.
These two factors, chain flexibility and packing density contribute to
the insolubility of polymers containing the aromatic ring. The limited

Table 9-1

COMPARISON OF VARIOUS PROPERTIES OF THE THREE HOMOPOLYMERS
AND THEIR COPOLYMERS PREPARED IN THIS STUDY

Structure		Conductivity (ohm cm)$^{-1}$	Sol. in Org. Solv.	Molecular Weight	Density	Stiffness	Stability
$(=C-\overset{H}{\underset{H}{C}}-N=N=)_x$	XVIII I$_2$ doped	$<10^{-8}$ $\sim 10^{-4}$	Yes Yes	High High	High High	High High	High High
$(=\langle\bigcirc\rangle-C=N-N=)_x$	XIX	$<10^{-8}$	No	Low	High	High	High
$(=C-C=N-N=)_x$	XX I$_2$ doped	$<10^{-8}$ $\sim 10^{-8}$	Yes Yes	High High	High High	Low Low	High High
$(=C-C=N-N=)_x-(=C-\langle\bigcirc\rangle-C=N-N=)_y$		$<10^{-8}$	No	Low	Low	Mixed	High
$(=C-C=N-N=)_x-(=C-C-C=N-N=)_y$		$<10^{-8}$	Yes	High	Low	Mixed	High
$(=C-C-C=N-N=)_x-(=C-\langle\bigcirc\rangle-C=N-N=)_y$		$<10^{-8}$	No	Low	Low	Mixed	High

solubility in turn prevents the polymerization of high molecular weight materials since the polymer precipitates at relatively low conversions. Again, almost all of the highly conducting polymers possess a high degree of intermolecular overlap and few if any are soluble in organic solvents because of their stiff long range conjugated backbones.

All of the compositions will take up iodine forming stable radical ions. The two that show increases in conductivity are those that are most capable of establishing intermolecular overlap necessary for electron transport. In the aromatic polyazine XIX the ions are too strongly localized, and the polymer of low molecular weight. The conjugated polymer XVIII has the stability, flexibility for solubility, and overlap necessary for electron transport. The addition of one more carbon atom into the backbone (polymer XX) is sufficient to strongly localize the radical ions and restrict transport.

While much work remains to be done before these materials are fully characterized, the available data indicates that their relative properties can be explained by the three factors that our model uses to explain conductivity: local electronic structure, flexibility, and packing density. This represents a successful departure from the conventional wisdom that treats conducting polymers as infinitely long, non-interacting, rigid, bandgap metals.

References

1.   S. Honig, Pure Appl. Chem. 15, 109 (1961).

2.   J. Vlanski et al., Poly. Plast. Tech. Eng. 17(2), 139 (1982).

3.   J. Mort and G. Pfister, Poly. Plast. Tech. Eng. 12, 89 (1970).

4.   Pittman et al., Macromolecules 12, 355, 541 (1979).

5.   W. Hertler, J. Org. Chem. 41(8), 1412 (1970).

6.   P. Grant and I. Batra, J. Synth. Metals 1, 193 (1980).

7.   C. Duke et al., Phys. Rev. B 18(10), 6717 (1978).

8. M. Pastraranu, M. Dumitriu, and T. Lixandru, European Polymer Journal <u>17</u>, 197 (1981).

9. G. F. D'Alelio, J. Macromol. Sci., Chem. <u>A2</u>(2), 237, 335 (1968).

10. G. Pulvermacher, Ber. Bunsenges. Phys. Chem. <u>26</u>, 2360 (1893).

11. T. C. Clarke, J. Raybolt, and G. Street, Polym. Prepr. <u>23</u>(1), 92 (1982).

12. H. Yamakawa, <u>Modern Therory of Polymer Solutions</u> (Harper and Row, New York, 1971).

10. Conclusions and Future Work

The conclusions and recommendations drawn from our study of electrically conducting polymers and the problem of spacecraft charging fall into three categories. The first deals directly with material-based approaches to minimizing or eliminating spacecraft charging problems. The second concerns the role that conducting and semiconducting polymers could play in new military systems. Finally, we address the basic scientific issues involved in preparing new materials and evaluating their properties.

Reduction of Spacecraft Charging Problems

The proposed solution to the charging of polymer dielectrics in spacecraft by high energy electrons involves the use of new or modified materials with semiconducting properties such that accumulated charge is drawn off before the breakdown potential is reached. The primary objective of this project was to find suitable materials for reducing spacecraft charging and/or identify means for developing such materials. We conclude that near-, intermediate-, and long-term approaches are available for obtaining these materials.

The immediate approach to the reduction of charging phenomena includes the laboratory investigation and testing in simulated spacecraft environments of polyvinyl carbazole (PVK) and pyropolymers produced from the partial pyrolysis of aromatic polyimides such as Kapton and polyacrylonitrile (PAN). Of the materials we surveyed, these have the best combination of electrical, thermal, and mechanical properties, radiation resistance, and availability.

The properties of polyvinyl carbazole have been studied in some detail, and its electrical properties can be controlled by processing and compounding techniques. It has adequate environmental resistance and is reasonably strong. The two properties that probably need to be improved are its toughness and strength retention at elevated temperatures. The softening point of PVK depends on its crystallinity, which is proportional to the isotactic content of chains. Research has shown that tacticity can be controlled through the use of various polymerization catalysts, and the development of a catalyst for producing 100% isotatic PVK would lead to a stronger, more thermally stable polymer. Two potential approaches to improving the toughness of PVK exist. The first is the development of copolymers of PVK and a tougher, energy-absorbing polymer analogous to the "toughened" epoxies and ABS resins. The second involves preparing a crosslinked PVK resin. By varying the polymer tacticity and crosslink density an optimum composition for spacecraft use could be identified.

The pyropolymers represent a second, immediate-term aprroach to the development of charge-resistant materials. PAN fibers and films are of relatively high strength, and graphitized PAN carbon fibers have exceptional strength-to-weight ratios. The pyrolysis product will probably not develop properties quite like those of graphite fiber, but should have reasonably good mechanical strength and environmental resistance. The properties of the pyrolysis product can be controlled, by varying the manufacturing process, to produce a material with electrical conductivities ranging over 5 to 6 orders of magnitude. Similarly, Kapton and related aromatic polyimides have excellent strength, thermal stability, and radiation resistance. The pyrolysis products span a wide range of conductivities. The major obstacle to the use of either of these materials is uncertainty in the mechanical properties (strength, modulus, and elongation at break) of the pyrolysis products.

A variety of intermediate-term solutions exist that involve the modification of polymer compositions that lack one or more of the properties required for specific space-based applications. For example, the polymer compositions currently used could be modified to include the stable, close-packed, delocalized, molecular ortibals that facilitate charge transport. Alternatively, the known semiconducting polymers could be chemically modified by various methods to raise their strength and thermal stabilty to the level required for specific missions.

This approach is more intermediate-term because it encompasses a very broad range of materials and applications. It can be best used in situations where materials are needed to fit a very specific set of physical requirements or where the use of a specific material is desired but a very specific performance objective, e.g., thermal stability or radiation resistance, must be improved.

The long-term opportunities are derived from an improved understanding of the phenomena of electrical conduction in organic polymers. The steps we have made in developing a general model of polymer conductivity have been confirmed by our identification and synthesis of new polymer compositions with electroactive properties. In addition to aiding the development of materials for other applications, an improved understanding of molecular structure, mechanical properties, and electrical conductivity will allow the design of completely new compounds tailored for specific applications rather than a series of modifications.

Systems Utilization of Electroactive Polymers

The use of charge-resistant dielectrics in spacecraft is just one example of how electroactive polymers can influence systems development. The development of polymer batteries is widely discussed but hindered by limitations in available materials. Our consideration of the chemical compositions that have electroactive properties and the problems encountered in the use of polyacetylene indicates that stable electrically conducting polymers capable of undergoing the same type of electrochemical redox reactions can be prepared. Because the electrochemical potential of these stable radical ion states is lower than that of oxygen or water, environmental degradation will be minimized or eliminated.

A mundane but universal problem is the development of rugged electrical connectors. The intrinsic elasticity and corrosion resistance of polymeric materials make them good candidates for use in connectors and switching devices. Materials with the necessary intrinsic electrical conductivity for these applications do not exist, but our model outlines specific structural properties that they should have.

Fundamental Scientific Questions

After surveying the properties of electroactive polymers, we conclude that there is a great need to better understand the mechanism of charge transport. Too many inconsistencies exist between the traditional band-gap approach and the observed experimental properties to rely on this model. There are indications that band-gap models are being discarded and the importance of localized charge states recognized, but considerable work needs to be done to incorporate these concepts in a complete model.

An important part of developing a better understanding of transport mechanisms is the development of relationships between polymer composition, structure, morphology, and electrical properties. The literature indicates that a great deal of diversity exists and that small changes in either composition or structure can produce dramatic changes in conductivity. The origin of these changes is unclear, and the use of band-gap model interpretations has not clarified the situation.

A major obstacle to understanding these structure-property relationships is that the majority of electroactive polymers are intractable and molecular characterization is impossible. We have shown that it is possible to prepare electroactive polymer systems that are processable and can be characterized. Continuing these investigations into the basic relationships between conductivity and structure could help clarify these basic scientific issues as well as aid in the development of materials for space-based use.

Appendix A

Material Characteristics
for Space-Based Applications

Tables A-1 through A-3 list the typical properties for materials used in space-based applications. Those qualities that may be considered as requirements for specific applications are noted with an asterisk (*). The specific values for each property include actual reported values or inferences based on the actual materials used in various applications. Sources include the reference lists of Section 3 and Appendix E.

Table A-1

MATERIAL CHARACTERISTICS FOR STRUCTURAL MATERIALS

Property	Primary Load Bearing	Secondary	Fiber Reinforced
Specific gravity, kg m^{-3}	1500	1200	1500
Tensile strength, MPa	100 *	20-50 *	10-100 *
Tensile modulus, MPa	5000 *	500-1000 *	3000 *
Shear strength, MPa	50	50	50
Compressive strength, MPa	100 *	50	100 *
Impact strength, J cm^{-1}	5	0.25	0.5
Flexure strength, MPa	500 *	100	100 *
Strain at failure, %	2-5 *	10	2-5 *
Thermal expansion coefficient, K^{-1}	$\leqslant 10^{-5}$	$<10^{-4}$	$<10^{-5}$ *
Thermal conductivity, J m^{-1} s^{-1} K^{-1}	1.5	1.5	1.5
Specific heat, kJ kg^{-1} K^{-1}	1.5	1.5	1.5

Table A-1 (Concluded)

Property	Primary Load Bearing	Secondary	Fiber Reinforced
Glass transition temperature, K	>500 *	350-500 *	>500 *
Dielectric constant	3	3	3
Electrical resistivity, ohm cm	10^{12} to 10^{18}	10^{12} to 10^{18}	10^{12} to 10^{18}
Dielectric strength	200	>100	200
Surface resistivity, ohm [per square]	10^{12} to 10^{16}	10^{12} to 10^{16}	10^{5} to 10^{16} depending on fiber content
Dissipation Factor	<0.003	<0.003	<0.003
Temperature range, K	Cryogenic to 500 *	150-350	Cryogenic to 500 *
Outgassing	TML 0.5% CVCM 0.05%	TML 0.5% CVCM 0.05%	TML 0.5% CVCM 0.05%
Ionizing radiation, Mrad	100 *	10-100	500 *
Heat distortion temperature, K	>500 *	350-500	>500 *

Table A-2

MATERIAL CHARACTERISTICS FOR DIELECTRICS, ADHESIVES, AND ELASTOMERS

Property	Dielectrics	Adhesives	Elastomers
Specific gravity kg m^{-3}	1200 to 1500	1200 to 1500	1200 to 1500
Tensile strength, MPa	20	50 *	10 to 50 *
Tensile modulus, MPa	500–1000	3000–5000 *	100 to 500
Shear strength, MPa	10	50 *	50 *
Compressive strength, MPa	50	50	25
Impact strength, J cm^{-1}	0.25	0.5	5 to 10
Flexure strength, MPa	50	100 *	25
Strain at failure, %	5	1 *	50 to 500 *
Thermal expansion coefficient, K^{-1}	10^{-4}	10^{-5} *	10^{-4}
Thermal conductivity, J m^{-1} s^{-1} K^{-1}	0.15	0.15	0.15
Specific heat kJ kg^{-1} K^{-1}	1.5	1.5	1.5

Table A-2 (Concluded)

Property	Dielectrics	Adhesives	Elastomers
Glass transition temperature, K	350 ✶	350-500 ✶	100 to 200 ✶
Dielectric constant	3-5 ✶	3	1 to 3
Electrical resistivity, ohm cm	$>10^{15}$ ✶	10^{12} to 10^{18}	10^{12} to 10^{18}
Dielectric strength, Volt/mil	200 ✶	100	100
Surface resistivity, ohm [per square]	$>10^{15}$	$>10^{12}$	$>10^{12}$
Dissipation Factor	0.003 ✶	0.003	0.003
Temperature range, K	150-350 ✶	150-350 ✶	200-300 ✶
Outgassing	TML 0.1% CVCM 0.005	TML 0.1% CVCM 0.005 ✶	TML 0.1% CVCM 0.005
Ionizing radiation, Mrad	10-50	1-10	1 to 10
Heat distortion temperature, K	350 ✶	350-500	200

Table A-3

MATERIAL CHARACTERISTICS FOR OPTICAL AND THERMAL CONTROL
MATERIALS AND FOR LUBRICANTS

Property	Optical	Thermal Control	Lubricants
Specific gravity, kg m^{-3}	1200–1500	1200–1500	1000–1200
Tensile strength, MPa	50–100	20–50 *	NA
Tensile modulus, MPa	500–1000	100	NA
Shear strength, MPa	10	10 *	NA
Compressive strength, MPa	50	10	NA
Impact strength, J cm^{-1}	0.35	0.25	NA
Flexure strength, MPa	100	50	NA
Strain at failure, %	1 *	5	NA
Thermal expansion coefficient, K^{-1}	10^{-5} *	10^{-5}	10^{-3} to 10^{-4}
Thermal conductivity, J m^{-1} s^{-1} K^{-1}	0.15	0.1 *	0.05
Specific heat, kJ kg^{-1} K^{-1}	1.5	1.0	1.0

NA = not applicable

Table A-3 (Concluded)

Property	Optical	Thermal Control	Lubricants
Glass transition, temperature, K	350 *	500 *	100 *
Dielectric constant	3	3	1 to 3
Electrical resistivity, ohm cm	10^{15}	10^{15}	$>10^{18}$
Dielectric strength	100	100	100-200
Surface resistivity, ohm [per square]	10^{12}	10^{12}	$>10^{18}$
Volume resistivity ohm cm	10^{15}	10^{15}	$>10^{18}$
Dissipation Factor	0.003	0.003	0.01
Temperature range, K	50-450 *	50-450 *	100-400 *
Outgassing	TML 0.1 CVCM 0.005 *	TML 0.5 CVCM 0.01 *	Large
Ionizing radiation, Mrad	100-1000 *	100-1000	1 to 10
Heat distortion temperature, K	350-400 *	350-400	NA

Appendix B
Standardized (ASTM) Test Methods

The standardized (ASTM) test methods used to evaluate
the material properties in Appendix A are listed in Table B-1.
Details of the test methods may be found in handbooks published
by the American Society for Testing and Materials, 1916 Race Street,
Philadelphia, PA 19103.

Table B-1

STANDARDIZED (ASTM) TEST METHODS

Property	ASTM No.
Specific gravity, kg M^{-3}	D792
Tensile strength, MPa	D638
Tensile modulus, MPa	D638
Shear strength, MPa	D732
Compressive strength, MPa	D695
Impact strength, J cm^{-1}	D256A
Flexure strength, MPa	D695
Strain at failure, %	D638
Thermal expansion coefficient, K^{-1}	D696
Thermal conductivity, $J\ m^{-1}\ s^{-1}\ K^{-1}$	C177
Specific, heat, kJ $kg^{-1}\ K^{-1}$	C351
Dielectric constant	D150
Dielectric strength, V mil^{-1}	D149
Surface resistivity, ohm	D257
Volume resistivity, ohm cm	D257
Outgassing	QRMO2T
Heat distortion temperature, K	D648

Appendix C

Relative Ratings of Structural Plastics

These ratings are taken from the "Space Materials Handbook" (reference 1, Section 3) and illustrate the range of materials that can be considered for specific applications.

Relative Rating	Application
A	Recommended for general use in interior and external structural applications, particularly involving high energy penetrating radiation and elevated temperatures.
B	Not preferred for primary structural use but may be used in place of A-rated materials if electrical or processing requirements dictate. Recommended for internal or external secondary structure and electrical or thermal applications.
C	Not recommended for high-load structural applications but satisfactory for all internal and external low-load or nonstructural uses and where electrical requirements dictate.
D	For structural and non-structural sandwich construction, internal insulation or electrical harness lacing applications.

STRUCTURAL PLASTICS

Material	Continuous Material Temperature Range (°F)	Radiation Damage Threshold (rad)	Ultraviolet and Vacuum Stability	Relative Rating	Design Remarks
Phenolic Resins, Glass Fabric Laminate	-300 to 500	8×10^8 to 8×10^9	Good	A	Preferred structural plastic. Thermosetting. Excellent resistance to temperatures of 100°F for short periods. Combined heat (to 900°F) and irradiation at doses up to 10^{10} rad produce no more degradation than does high temperature alone. Excellent mechanical properties throughout operating temperature range if properly postcured.
Epoxy Resin, Glass Fabric Laminate (Low Pressure) and Sheet (High Pressure)	-300 to 250	2×10^9 to 5×10^9	Good	A	Preferred structural plastic. Thermosetting. Undamaged in significant mechanical and electrical properties at dose of 10^9 rad, except impact strength which decreases approximately 30%. Best mechanical properties of any reinforced plastic type at ambient temperatures but strength decreases rapidly at elevated temperatures.

Material	Continuous Material Temperature Range (°F)	Radiation Damage Threshold (rad)	Ultraviolet and Vacuum Stability	Relative Rating	Design Remarks
Silicone Resin, Glass Fabric Laminate (Low Pressure) and Laminate (Low Pressure Prepreg)	-300 to 500	10^9 to 5×10^9	Good	B	Excellent elevated temperature electrical properties. Good mechanical properties at temperatures of 750°F for short exposures.
Polyester Resin, Glass Fiber Laminate, Low Pressure	-100 to 200 -100 to 300	2×10^9 to 5×10^8 8×10^9 to 2×10^{10}	Good	B	Thermosetting. Laminate has good electrical and structural properties. General-purpose structural laminate.
Phenolic Resin, Glass Fiber Molding	-300 to 450	8×10^8 to 8×10^9	Good	C	General-purpose moldings. Good strength and temperature resistance. Not for primary structural applications.
Silicone Resin, Glass Fiber Molding	-300 to 500	10^9 to 5×10^9	Good	C	Not for primary structural applications.
Polyester Resin, Glass Fiber Molding	-100 to 180	2×10^9 to 5×10^9	Good	C	Not for primary structural applications. General-purpose non-structural moldings. Thermosetting. Good physical and electrical properties.
Melamine Resin, Glass Fiber Molding	-65 to 350	10^8 to 10^9	Good	C	Not for primary structural applications. Excellent electrical properties, moderate temperature resistance. Thermosetting.
Modified epoxy, Phenolic-Glass Fiber (heat resistant epoxy)	-300 to 500	5×10^9	Good	B	For general use in interior and exterior structural applications particularly where moderate to high temperatures and radiation exposures are encountered; excellent mechanical strength properties and good stability to vacuum and UV radiation.

Material	Continuous Material Temperature Range (°F)	Radiation Damage Threshold (rad)	Ultraviolet and Vacuum Stability	Relative Rating	Design Remarks
Films					
Polyimide (H-Film)	-423 to 600	5×10^8	Good	B	Preferred materials for internal and external semi-structural and related applications (e.g., inflatable space structures) particularly where long-term resistance to UV radiation is required and moderate temperatures and radiation doses are encountered; good vacuum stability and mechanical strength properties. Polymer films are generally aluminized on one side in most spacecraft applications.
Polyvinylfluoride (Teslar or Tedlar)	-100 to 250	10^8	Good	B	
Polyester Film (Mylar)	-100 to 300 (Low load applications)	5×10^7 to 1×10^8	Fair to poor UV stability unless metallized. Good in vacuum.	C	Inflatable semistructural film. Thermosetting type. For space-craft, antennas, etc. Low vapor permeability. Can be rigidized in space with foam-in-place materials and techniques. Aluminizing surface improves UV stability in space.
Polyethylene (High Density)	-320 to 225	10^8	Poor	D	Not preferred for long-term exposures to UV radiation unless surface is metallized, but may be used for short missions internally or externally at low to moderate exposures and moderate temperatures; good vacuum stability. Note that radiation resistance of Teflon (FEP and TFE) is improved about two orders of magnitude in vacuum or inert atmosphere.
Polypropylene	-423 to 275	10^8	Fair	D	
Kel-F	-423 to 350	10^6	Good	C	
Teflon (TFE)	-423 to 500	5×10^4	Good	C	
Teflon (FEP)	-423 to 450	5×10^5	Good	C	

Material	Continuous Material Temperature Range (°F)	Radiation Damage Threshold (rad)	Ultraviolet and Vacuum Stability	Relative Rating	Design Remarks
Polyethylene (Low Density)	-320 to 180	10^8	Poor	D	Not recommended for external applications because of UV radiation susceptibility; may be used internally in applications where low-to-moderate temperatures are encountered.
Polyvinyl Chloride (PVC)	-100 to 165	10^8	Poor	C	
Polyurethane, Rigid Foam	-300 to 350	10^8 to 10^9	Good	D	Foamed sandwiches do not degrade in flexure and compression up to 10^9 rads. Vibration damping. Good electrical properties. Density range: 2 to 20 lb/ft^3 (closed cell). Thermoplastic.
Polystyrene, Rigid Foam	-100 to 165	8×10^9 to 4×10^9	Good	D	Thermal insulation. Sandwich construction for low load applications. Density range: 2 to 6 lb/ft^3. Thermoplastic.
Foams (Closed Cell, Rigid Thermoset Types)					
Polyether	-423 to 300	10^9	Not applicable; protected by sandwich skin	D	General-purpose foamed sandwich and honeycomb materials for internal spacecraft applications.
Phenolic	-423 to 450	10^9		D	
Epoxy	-423 to 250	10^9		D	
Silicone	-423 to 500	10^8		D	
Fibers					
HT-1	-423 to 600	5×10^8	Good	D	General-purpose use for electrical harness lacing and other non-structural applications. Dacron has slightly better thermal-vacuum stability than nylon.
Dacron (Polyester)	-423 to 300	10^8	Good	D	
Nylon	-320 to 300	5×10^7	Fair	D	

Appendix D
Relative Ratings of Dielectric Materials

These ratings are taken from the "Space Materials Handbook" (reference 1, Section 3) and illustrate the range of materials that can be considered for specific applications.

Relative Rating	Application
A	Preferred materials for insulation and dielectric applications with stable properties in severe environments.
B	Satisfactory material with application limited by initial properties or proper degradation in radiation, or thermal environments.
C	Use of such material limited to specific application unless shielded by protection of vehicle skin or container or some loss of certain properties can be tolereated.

Generally "A" rated materials possess dielectric strengths in excess of 200 V/mil and good environmental resistance. "B" rated materials have high ($\sim$200 V/mil) dielectric strengths but lack environmental resistance. "C" rated materials have relatively low (150-175 V/mil) dielectric strengths and are limited to internal applications.

Material	Continuous Material Temperature Range (°F)	Radiation Damage Threshold (rad)	Stability-Space UV and Vacuum	Relative Rating	Remarks
Electrical Insulation					
Polyethylene Expanded Tubing, Heat Shrinkable Types 1, 2, and 3	to 215	2×10^7 to 10^8	Fair	A	Type 1 - Heat-shrinkable jacket insulation for equipment. Type 2 - Material for repair of jacket defects. Type 3 - Dual-wall shrinkable material for embedding and parts ($\sim$75%). Excellent electrical properties, for wire connector insulation and membrane insulation.
Modified Neoprene Rubber Tubing Heat Shrinkable, Flexible	-65 to 250	3×10^6 to 3×10^8	Fair	C	For repair of jackets where oil resistance is needed; softens in benzene group solvents and in ketones.
Polytetrafluoroethylene Coated Tape, Glass Braid Lacing	to 500	10^6	Good	B	Teflon coating limits radiation resistance, excellent electrical properties at elevated temperature.
Vinyl Coated Tape, Glass Braid Lacing and Typing	to 215	2×10^7 to 5×10^8			Excellent electrical properties.
Irradiated Modified Polyolefin Insulation for Vehicle Wiring	-65 to 275	5×10^8	Fair	A	Used for vehicle wiring and all flight items, thermally stabilized, and irradiated modified polyolefin insulated wire.
Polyamide (Nylon)	-65 to 275	5×10^8	Good	C	Electrical insulation.
Polyester (Mylar), Glass Filled	-70 to 300	2×10^9 to 5×10^9	Good	B	Electrical insulation.
Polytetrafluoroethylene (Teflon)	to 400	10^5 (air) to 10^6 (vacuum)	Good	B	Special applications requiring unique electrical properties or heat resistance, where radiation total dose is low.

Material	Continuous Material Temperature Range (°F)	Radiation Damage Threshold (rad)	Stability- Space UV and Vacuum	Relative Rating	Remarks
Natural Rubber (Anti-Rad Type)	-40 to 225	2×10^7 to 2×10^8	Poor	B	Elastomer for vibration damping; wire insulation and sheathing.
Neoprene Rubber (Polychloroprene)	-65 to 250	3×10^6 to 2×10^8	Fair	C	Elastomeric cable sheathing, molded parts, bumpers, and shock absorbers, jacketing and sheathing only, not insulation.
Synthetic Rubber (Buna-N)	-40 to 250	10^7	Fair	C	Cable jacket and tubing material; wire insulation.
Neoprene Sheath, Buna-N insulated	-40 to 250	6×10^6 to 3×10^7	Fair	C	Cable sheath; wire insulation.
Butyl Rubber	-65 to 250	10^6 to 10^7	Good	C	Elastomer for vibration damping; wire insulation or sheathing.
Polyvinyl Formal (Formvar)	-75 to 325	2×10^7 to 3×10^8	Good	B	Magnet wire insulation; solar array application.
Polyethylene	-70 to 250	2×10^7 to $2 \times 1 10^8$	Fair	A	Embrittled in UV radiation.
Polyester (Mylar)	-100 to 300	5×10^7 to 5×10^8	Fair in UV unless metallized. Good in vacuum.	B	Electrical insulation in manufacture of cables. Thermosetting.
Polyvinyl Chloride	-500 to 200	2×10^7 to 2×10^8	Poor	C	Electrical insulation and vibration protection. Plasticized tyges outgas and embrittle above 10^8 rad. Liberties HCl above 10^7 rad.

Material	Continuous Material Temperature Range (°F)	Radiation Damage Threshold (rad)	Stability- Space UV and Vacuum	Relative Rating	Remarks
Kel-F (Monochlorotrifluorethylene) AMS 3650	-80 to 300	3×10^6 to 2×10^7	Fair	C	Electrical insulation, thermo-plastic elastomer; embrittled in UV radiation; may evolve corrosive gases.
Tetrafluorethylene (Teflon)	-150 to 500	10^5 air to 10^6 vacuum	Good	B	Electrical insulation, thermo-plastic; low radiation resistance, releasing corrosive fluoride gases at higher doses.
Polymethyl Methacrylates	-80 to 190	1×10^7	Good in vacuum; fiar in UV	C	Magnet wire insluation; thin coating.
Circuit Boards and Other Electrical Laminates					
Copper-Clad Epoxy, Glass Laminates	-60 to 300	5×10^9	Good	A	Printed circuit board.
Epoxy-Glass Laminate	-65 to 300	5×10^9	Good	A	Material in thickness of 7 mils up to 0.25 in.
Epoxy-Glass Laminate	-65 to 250	5×10^9	Good	A	Satisfactory where used within temperature limits; antenna window and related uses.

Appendix E

Properties of Commercial and Experimental Polymers Relevant to Space-Based Uses

The following is a list of materials that are used in spacecraft or are candidates for use. The properties of each material were evaluated based on the materials characteristics listed in Appendix A. The number in parentheses corresponds to the reference listed in the Reference List at the end of this appendix. Values without a reference are estimates.

For some properties where ASTM standards are not used (like radiation resistance) or where test conditions vary (notch size in impact test or stress in heat distortion temperature) direct comparisons between entries may not be meaningful and it may be necessary to consult original reference. Entries for a class, e.g., epoxies give typical values or a range of reported values.

147

	Polycarbonate Poly(oxycarbonyloxy-1,4-phenylene isopropylidene-1,4-phenylene) (Lexan)		Poly(tetrafluoromethylene) (Teflon)		Acetal Copolymer Poly(oxymethylene-coethylene) (Celcon)	
Specific gravity Kg m^{-3}	1200	(1)	2150	(1)	1410	(1)
Tensile strength MPa	90	(1)	25	(1)	73	(1)
Tensile modulus MPa	6000	(1)	420	(1)	3500	(1)
Shear strength MPa	72.4	(2)	10		50	
Compressive strength MPa	124	(3)	10-16	(3)	110	(3)
Impact strength J cm^{-1}	7.95	(4)	1.6 notched	(6)	1.78 notched	(1)
Flexure strength MPa	158	(3)	50		90-100	(6)
Strain at failure %	110	(4)	300-450	(3)	40	(3)
Thermal expansion coefficient K^{-1}	3 x 10^{-5}	(1)	1 x 10^{-4}	(1)	1.1 x 10^{-4}	(1)
Thermal conductivity J m^{-1}s^{-1}K^{-1}	0.160	(1)	0.25	(1)	0.31	(1)
Specific heat KJ Kg^{-1}K^{-1}	1.1	(1)	1.0	(1)	1.46	(1)

	Polycarbonate Poly(oxycarbonyloxy-1,4-phenylene isopropylidene-1,4-phenylene) (Lexan)		Poly(tetrafluoroethylene) (Teflon)		Acetal Copolymer Poly(oxymethylene-coethylene) (Celcon)	
Glass transition temperature K	418	(5)	160, 400; conflicting data	(5)	~200	
Dielectric constant	3.3 @ 50 Hz	(1)	2.1 @ 60 Hz	(1)	4 @ 50 Hz	(1)
Electrical resistivity Ωcm	>10^{17}		8 x 10^{16}		>10^{15}	
Dielectric strength	187	(2)	189	(1)	700	(1)
Surface resistivity Ω [per square]	10^{15}		10^{13}		1.3 x 10^{16}	(3)
Dissipation Factor	9 x 10^{-4} @ 50 Hz	(1)	2.5 x 10^{-3} @ 60 Hz	(1)	1.1 x 10^{-3} @ 50 Hz	(1)
Temperature range K	<418		Cryogenic to 533	(1)	233-373	(1)
Outgassing	TML 0.15% CVCM 0%	(1)	TML 0.05% CVCM 0.0%	(1)	TML 0.4% CVCM 0.02%	(1)
Ionizing radiation MRad	10 (mechanical properties)		10 (vacuum) 0.1 (air) (damage threshold)	(7)	5 (mechanical properties)	(1)
Heat distortion temperature K	419 @ 1.82 MPa		294 @ 0.45 MPa		383-400 @ 1.82 MPa	(6)

	Acetal Poly(oxymethylene) (Delrin)		PET Poly(ethylene- terephthalate)		PVF_2 Poly(vinylidine- fluoride) (Kynar)	
Specific gravity $Kg\ m^{-3}$	1420	(3)	1400	(1)	1800	(1)
Tensile strength MPa	69	(3)	21	(1)	56	(1)
Tensile modulus MPa	3590	(3)	2792	(3)	7000	(1)
Shear strength MPa	53	(8)	50		55	(3)
Compressive strength MPa	124	(3)	100		75	
Impact strength $J\ cm^{-1}$	0.7 notched 10.6	(8)	0.5 notched		2.0 notched	(6)
Flexure strength MPa	97	(3)	110	(3)	60	(3)
Strain at failure %	40	(3)	70	(3)	25–500	(3)
Thermal expansion coefficient K^{-1}	2.61×10^{-5}	(8)	3×10^{-5}	(1)	2.62×10^{-5}	(3)
Thermal conductivity $J\ m^{-1}s^{-1}K^{-1}$	0.37	(3)	0.61	(1)	0.24	(3)
Specific heat $KJ\ Kg^{-1}K^{-1}$	1.47	(3)	1.30	(1)	1.38	(3)

	Acetal Poly(oxymethylene) (Delrin)		PET Poly(ethylene- terephthalate)		PVF$_2$ Poly(vinylidine- fluoride) (Kynar)	
Glass transition temperature K	243	(5)	342	(5)	233	(5)
Dielectric constant	3.7 @ 50 to 10^6 Hz	(3)	3.2 @ 50 Hz	(1)	8.4 @ 10^2 Hz 7.7 @ 10^3 Hz 6.4 @ 10^6 Hz	(3)
Electrical resistivity Ωcm	10^{15}		6.4 x 10^9 - 1 x 10^{18} (humidity sensitive)	(3)	2 x 10^{15}	(1)
Dielectric strength	197	(3)	280	(1)	300	(1)
Surface resistivity Ω [per square]	1 x 10^{15}	(3)	6 x 10^{14}	(3)	10^{12}	
Dissipation Factor	0.003 @ 50- 1000 Hz 0.0005 @ 10^6 Ha	(3)	2 x 10^{-3} @ 50 Hz	(1)	0.049 @ 10^2 Hz 0.018 @ 10^3 Hz 0.17 @ 10^6 Hz	(3)
Temperature range K	100-300		100-450	(1)	220-450	(1)
Outgassing	TML 0.2% CVCM 0.05%		TML 0.3% CVCM 0.02%	(1)	TML 0.36% CVCM 0.07%	(1)
Ionizing radiation MRad	5		10 (mechanical properties)	(1)	10 (air) 50 (max)	(7)
Heat distortion temperature K	283-300 @ 1.82 MPa	(6)	350		425 @ 0.45 MPa	(6)

	Polyimide (Kapton H)		Polyurethane (Solithane 113)		Silicones	
Specific gravity Kg m^{-3}	1420	(1)	1073	(1)	1080–1880	(1)
Tensile strength MPa	17	(1)	2.85		3.40–6.60	
Tensile modulus MPa	350	(1)	200	(1)	150–250	(1)
Shear strength MPa	75		10		10–50	
Compressive strength MPa	16.5	(5)	20		96.5	(2)
Impact strength J cm^{-1}	7.5		2.5		3.45 notched	(6)
Flexure strength MPa	150		80		79.2	(2)
Strain at failure %	5–8	(5)	400–450	(3)	100–500	
Thermal expansion coefficient K^{-1}	2×10^{-5}	(1)	1.26×10^{-4}	(1)	3.0×10^{-4}	(1)
Thermal conductivity J m^{-1}s^{-1}K^{-1}	0.155	(1)	0.20	(1)	0.146–0.30	(1)
Specific heat KJ Kg^{-1}K^{-1}	1.09	(1)	1.80	(1)	1.5	

	Polyimide (Kapton H)		Polyurethane (Solithane 113)		Silicones	
Glass transition temperature K	583	(5)	213-412	(5)	200-500	
Dielectric constant	3.5 @ 10^3 Hz	(1)	3.0 @ 10^2 Hz 2.9 @ 10^3 Hz 2.8 @ 10^6 Hz	(1)	4.2 @ 60 Hz 2.75 @ 110 Hz	(1)
Electrical resistivity Ωcm	$10^{13} - 10^{15}$	(23)	2.5×10^{14}		$7 \times 10^{13} - 10^{15}$	(1)
Dielectric strength	325	(1)	150	(1)	196	(1)
Surface resistivity Ω [per square]	10^{15}		1×10^{15}	(1)	10^{15}	
Dissipation Factor	0.003 @ 10^3 Hz	(1)	0.032 @ 10^2 Hz 0.033 @ 10^3 Hz 0.012 @ 10^6 Hz	(1)	0.0011 @ 10^6 Hz	(2)
Temperature range K	Cryogenic to 575	(1)	Cryogenic to 450	(7)	230-355	(15)
Outgassing	TML 1.3% CVCM 0.02%	(1)	TML 0.4% CVCM 0.03	(1)	TML 0.18% CVCM 0.02	(1)
Ionizing radiation MRad	500	(1)	1-10	(7)	5-200	(15)
Heat distortion temperature K	500		250		755	(6)

	Polyolefin Poly(ethylene-co-propylene) (Thermofit)		Ionomer Poly(ethylene-co-acrylic acid) metal salt (Surlyn)		Nylon Poly(Hexamethylene adipamide) (Nylon 66)	
Specific gravity Kg m^{-3}	1300	(1)	936–966	(6)	1060–1150	(8)
Tensile strength MPa	14	(1)	26.9	(3)	67.1	(3)
Tensile modulus MPa	670	(1)	140–410	(5)	2792	(3)
Shear strength MPa	20		100		58–66	(8)
Compressive strength MPa	55	(3)	50		101	(3)
Impact strength J cm^{-1}	0.5 notched		5.83 notched	(6)	0.32–0.53 notched	(8)
Flexure strength MPa	50		91.3	(3)	91.3	(3)
Strain at failure %	40	(3)	132	(3)	132	(3)
Thermal expansion coefficient K^{-1}	36 x 10^{-5}	(3)	2.56 x 10^{-5}	(8)	2.56 x 10^{-5}	(8)
Thermal conductivity J $m^{-1}s^{-1}K^{-1}$	0.17	(3)	0.24	(3)	0.21	(3)
Specific heat KJ $Kg^{-1}K^{-1}$	1.97	(3)	2.26	(3)	1.47	(3)

	Polyolefin Poly(ethylene-co-propylene) (Thermofit)		Ionomer Poly(ethylene-co-acrylic acid) metal acid (Surlyn)		Nylon Poly(Hexamethylene adipamide) (Nylon 66)	
Glass transition temperature K	175		450–500		313–384	
Dielectric constant	2.1 @ 10^2 – 10^6 Hz	(3)	2.4 @ 10^6 Hz	(3)	4.55 @ 10^2 Hz 4.32 @ 10^3 Hz 3.77 @ 10^6 Hz	(3)
Electrical resistivity Ωcm	1×10^{14}	(1)	10^{18}		5×10^6 – 3×10^{17} (humidity sensitive)	(9)
Dielectric strength	197	(1)	433	(3)	264	(3)
Surface resistivity Ω [per square]	10^{16}		10^{16}		5×10^{13}	(3)
Dissipation Factor	0.0005 @ 10^6 Hz		0.001 @ 10^2 Hz 0.002 @ 10^6 Hz	(3)	0.019 @ 10^2 Hz 0.029 @ 10^3 Hz 0.048 @ 10^6 Hz	(3)
Temperature range K	220–410		100–600		77–575	(7)
Outgassing	TML 0.8% CVCM 0.09%	(1)	TML 0.5% CVCM 0.05%		TML 1% CVCM 0.05%	
Ionizing radiation MRad	20–50		1–5		0.9 continuous 5 max tolerable	(7)
Heat distortion temperature K	34 @ 0.45 MPa	(6)	400		330 @ 1.82 MPa	

	Poly(phenylene sulfide)		Poly(p-xylene)		Polyphenylene oxide (Noryl)	
Specific gravity Kg m^{-3}	1340	(8)	1200		1060	(8)
Tensile strength MPa	69	(3)	62	(3)	66	(8)
Tensile modulus MPa	2500		1000		2445	(8)
Shear strength MPa	50		35		72.4	(8)
Compressive strength MPa	113.6	(3)	75		100	
Impact strength J cm^{-1}	0.159 notched 1.6-2.1	(8)	0.2 notched		0.6 notched 6.3	(8)
Flexure strength MPa	117.1	(3)	100		103	(8)
Strain at failure %	0.5	(3)	30	(3)	20	(8)
Thermal expansion coefficient K^{-1}	1.67×10^{-5}	(8)	1.5×10^{-5}		1.83×10^{-5}	(8)
Thermal conductivity J $m^{-1}s^{-1}K^{-1}$	0.288	(8)	0.13	(3)	0.216	(8)
Specific heat KJ $Kg^{-1}K^{-1}$	1.7		1.3		1.4	

	Poly(phenylene sulfide)		Poly(p-xylene)		Polyphenylene oxide (Noryl)	
Glass transition temperature K	358	(12)	400		482	(8)
Dielectric constant	5.4 @ 10^3 Hz 6.1 @ 10^6 Hz	(3)	2.7 @ 10^3 Hz	(3)	2.65 @ 60 Hz 2.65 @ 10^6 Hz	(8)
Electrical resistivity Ωcm	10^{15}-10^{18} undoped 10^0-10^3 doped	(12)	10^{16}		10^{17}	
Dielectric strength	183	(8)	275	(8)	216	(8)
Surface resistivity Ω [per square]	10^{11}		10^{14}		10^{16}	(8)
Dissipation Factor	0.012 @ 10^3 Hz 0.068 @ 10^6 Hz	(8)	0.0002 @ 10^3 Hz	(8)	0.0005 @ 60 Hz 0.001 @ 10^6 Hz	(8)
Temperature range K	200-395	(7)	100-400		100-500	
Outgassing	TML 0.5% CVCM 0.05%		TML 0.5% CVCM 0.05%		TML 0.5% CVCM 0.05%	
Ionizing radiation MRad	1-100 (air)	(7)	50		25	
Heat distortion temperature K	350		375		450	

	Epoxy		Crosslinked Polystyrene		Polystyrene	
Specific gravity Kg m^{-3}	1200	(1)	1050	(1)	1050	(3)
Tensile strength MPa	40-92		50	(1)	26.5	(3)
Tensile modulus MPa	2550-9500	(1)	2500-5000		2645	(3)
Shear strength MPa	50		50		25	
Compressive strength MPa	124	(3)	100		71.5	(3)
Impact strength J cm^{-1}	0.265 notched	(1)	0.25 notched		0.5-8.0 notched	(6)
Flexure strength MPa	110	(3)	100-200		130	(3)
Strain at failure %	3-7	(3)	5		18-22	(3)
Thermal expansion coefficient K^{-1}	6 x 10^{-5}	(1)	7 x 10^{-5}	(1)	2.16 x 10^{-5}	
Thermal conductivity J m^{-1}s^{-1}K^{-1}	0.170	(1)	0.146	(1)	0.144	(6)
Specific heat KJ Kg^{-1}K^{-1}	1.2-1.5		1.39		1.39	(6)

	Epoxy		Crosslinked polystyrene		Polystyrene	
Glass transition temperature K	500-300		375		373	(5)
Dielectric constant	3.4-3.9	(1)	2.53	(1)	2.6 @ 10^2 – 10^6 Hz	(3)
Electrical resistivity Ωcm	4×10^{16}	(1)	$>10^{16}$		$>10^{17}$	
Dielectric strength	200	(1)	197	(1)	217	(3)
Surface resistivity Ω [per square]	5.7×10^{12}	(6)	10^{15}		10^{15}	
Dissipation Factor	0.003 @ 10^2 Hz 0.006 @ 10^3 Hz 0.03 @ 10^6 Hz	(3)	0.0001 @ 10^6 Hz	(1)	0.0002 @ 10^2 – 10^6 Hz	(3)
Temperature range K	80-400	(7)	200-375	(1)	200-350	
Outgassing	TML 0.4% CVCM 0.07%	(1)	TML 0.2% CVCM 0.01%	(1)	TML 0.2% CVCM 0.01%	
Ionizing radiation MRad	10-100	(1) (7)	1000	(1)	100	
Heat distortion temperature K	400-500		350		370	(6)

	Poly(acrylonitrile) (PAN)		Polyamideimide		Poly(vinylidine fluoride-co-hexafluoro propylene) (Viton)	
Specific gravity Kg m^{-3}	1200		1400–1450	(2)	1850	(1)
Tensile strength MPa	83.3	(3)	50		15	(1)
Tensile modulus MPa	1000		500		6.5	(1)
Shear strength MPa	75		185.3	(3)	25	
Compressive strength MPa	50		220	(3)	20	
Impact strength J cm^{-1}	0.5 notched		1.32 notched 7.42 unnotched	(8)	2 notched	
Flexure strength MPa	125		211.4	(3)	75	
Strain at failure %	5–10		12	(3)	75	
Thermal expansion coefficient K^{-1}	10^{-4}		1.11 x 10^{-5}	(3)	5 x 10^{-4}	
Thermal conductivity J m^{-1}s^{-1}K^{-1}	0.25	(3)	0.24	(3)	0.25	
Specific heat KJ Kg^{-1}K^{-1}	1.26		1.1		1.0	

	Poly(acrylonitrile) (PAN)	Polyamideimide	Poly(vinylidine fluoride-co-hexafluoro propylene) (Viton)
Glass transition temperature K	370 (5)	300	200
Dielectric constant	4	3.5 @ 10^3 Hz (3)	2
Electrical resistivity Ωcm	$10^{-2} - 10^7$ (10)	1×10^{17}	2×10^{13} (1)
Dielectric strength	200	236 (3)	175
Surface resistivity Ω [per square]	10^{15}	1×10^{17} (3)	2×10^{13}
Dissipation Factor	0.001	0.001 @ 10^3 Hz (3) 0.009 @ 10^6 Hz	0.0005
Temperature range K	100-400	200-350	200-500 (1)
Outgassing	TML 0.2% CVCM 0.05	TML 0.7% CVCM 0.07	TML 0.5% (1) CVCM 0.02%
Ionizing radiation MRad	5	10	10 (1)
Heat distortion temperature K	320	400	250

	Polyethylene		Polypropylene	
Specific gravity Kg m^{-3}	910–965	(6)	903	(3)
Tensile strength MPa	4–38	(5)	34.5	(3)
Tensile modulus MPa	137	(3)	138	(3)
Shear strength MPa	20		20	
Compressive strength MPa	19–25	(5)	38–55	(5)
Impact strength J cm^{-1}	8.48 0.1–0.4 notched	(6)	0.2–7.95	(6)
Flexure strength MPa	33–48	(5)	54.4	(3)
Strain at failure %	500	(3)	250	
Thermal expansion coefficient K^{-1}	4.63×10^{-5}	(3)	2.78×10^{-5}	(3)
Thermal conductivity J m^{-1}s^{-1}K^{-1}	0.33	(3)	0.18	(3)
Specific heat KJ Kg^{-1}K^{-1}	230	(6)	260–270	(3)

	Polyethylene		Polypropylene	
Glass transition temperature K	148	(3)	260–270	(3)
Dielectric constant	$2.3 @ 10^6$ Hz	(3)	$2.3 @ 10^2 -$ 10^6 Hz	(3)
Electrical resistivity Ωcm	10^{16}		10^{16}	
Dielectric strength	276	(3)	256	(3)
Surface resistivity Ω [per square]	10^{14}		10^{14}	
Dissipation Factor	$0.0005 @ 10^6$ Hz	(3)	$0.0005 @ 10^3 -$ 10^6 Hz	(3)
Temperature range K	150–355	(7)	150–300	
Outgassing	TML 0.2% CVCM 0.05%		TML 0.2% CVCM 0.05%	
Ionizing radiation MRad	20–90	(7)	20–50	
Heat distortion temperature K	315–355 @ 1.82 MPa	(6)	320–350 @ 1.82 MPa	(6)

	Polypyrrole		Polyvinyl Pyridine	Polyvinylcarbazole
Specific gravity Kg m^{-3}	1480	(16)	1400	1250
Tensile strength MPa	25		50	75
Tensile modulus MPa	500		1000	1000
Shear strength MPa	25		75	75
Compressive strength MPa	25		50	75
Impact strength J cm^{-1}	0.25 notched		0.5 · notched	0.5 notched
Flexure strength MPa	100		150	200
Strain at failure %	5		5^{-10}	2-5
Thermal expansion coefficient K^{-1}	1×10^{-4}		2×10^{-4}	1×10^{-4}
Thermal conductivity J $m^{-1}s^{-1}K^{-1}$	0.35		0.15	0.20
Specific heat KJ $Kg^{-1}K^{-1}$	1.2		1.2	1.2

	Polypyrrole	Polyvinyl Pyridine	Polyvinylcarbazole
Glass transition temperature K	dec.	377-415 (5)	4235
Dielectric constant	3	3	3
Electrical resistivity Ωcm	2×10^2 - (16) 1×10^{-2}	10^{15} (17) 4×10^3 TCNQ complex 10^4 - 10^9 iodine	10^{12}
Dielectric strength	100	150	250
Surface resistivity Ω	10^2	10^{13}	2.2×10^{13} (18) humidity sensitive
Dissipation Factor	0.003	0.001	0.001
Temperature range K	200-350	100-300	100-300
Outgassing	Reversible wt (16) loss @ 573 K	TML 0.5 CVCM 0.03	TML 0.5 CVCM 0.03
Ionizing radiation MRad	5-10	1-5	5-10
Heat distortion temperature K	375	375	400

	Poly(benzobisthiazole) PBT		Polyphthalocyanines	Polyacylene quinone radical polymers
Specific gravity Kg m^{-3}	1480		1300	1250
Tensile strength MPa	2400	(25)	100	75
Tensile modulus MPa	250,000	(25)	5000	4000
Shear strength MPa	100		100	75
Compressive strength MPa	100		100	50
Impact strength J cm^{-1}	0.75 notched		0.5 notched	0.5 notched
Flexure strength MPa	200		200	150
Strain at failure %	1.5	(25)	2-5	2-5
Thermal expansion coefficient K^{-1}	1 x 10^{-5}		5 x 10^{-5}	5 x 10^{-5}
Thermal conductivity J m^{-1}s^{-1}K^{-1}	0.35		0.25	0.20
Specific heat KJ Kg^{-1}K^{-1}	1.5		1.4	1.2

	Poly(benzobisthiazole) PBT	Polyphthalocyanines	Polyacylene quinone radical polymers
Glass transition temperature K	dec.	dec.	dec.
Dielectric constant	3.5	16-1300 (27)	50-6000 (28)
Electrical resistivity Ωcm	10^{12}-10^{17} (26)	7-3×10^6 (27)	10^1-10^{11} (28)
Dielectric strength	2500	2000	3000
Surface resistivity Ω [per square]	10^{12}	10^4	10^1-10^{10}
Dissipation Factor	0.003	0.005	0.005
Temperature range K	Cryogenic to 600	Cryogenic to 500	100-500
Outgassing	TML 0.5% CVCM 0.05%	TML 0.5% CVCM 0.05%	TML 0.5% CVCM 0.05%
Ionizing radiation MRad	10-50	10-50	1-5
Heat distortion temperature K	dec.	dec.	dec.

	SN$_x$	Pyrolyzed Kapton		Emeralidines
Specific gravity Kg m^{-3}	1250	1650	(23)	1200
Tensile strength MPa	10	5		2.5
Tensile modulus MPa	500	300		250
Shear strength MPa	25	75		50
Compressive strength MPa	25	50		50
Impact strength J cm^{-1}	0.25 notched	0.25 notched		0.25 notched
Flexure strength MPa	75	100		100
Strain at failure %	1-2	2-5		5-7
Thermal expansion coefficient K^{-1}	7 x 10^{-5}	1 x 10^{-5}		1 x 10^4
Thermal conductivity J m^{-1}s^{-1}K^{-1}	0.25	0.146		0.20
Specific heat KJ Kg^{-1}K^{-1}	1.5	1.0		1.2

	SN$_x$	Pyrolyzed Kapton	Emeralidines
Glass transition temperature K	dec.	dec.	dec.
Dielectric constant	3.5	2	1.5
Electrical resistivity Ωcm	10^3-10^{14} (22)	10^3-10^{-1} (23)	10^4-10^{-1} (24)
Dielectric strength	2000	2500	1000
Surface resistivity Ω [per square]	10^{12}	10^3-10^1	10^2
Dissipation Factor	0.005	0.003	0.001
Temperature range K	Cryogenic to 350	Cryogenic to 500	100-350
Outgassing	TML 0.05 CVCM 0.005	TML 0.1 CVCM 0.05	TML 0.05 CVCM 0.01
Ionizing radiation MRad	1	100-1000	1-10
Heat distortion temperature K	dec.	dec.	dec.

	DA Complex of Polyesters and Polycarbonates		TTF:TCNQ Containing Polyurethane		Metal Ion Coordination Polymers
Specific gravity Kg m^{-3}	1200–1400		1100		1200–2000
Tensile strength MPa	34.5	(29)	10–40	(30)	10–20
Tensile modulus MPa	2480	(29)	500–1000	(30)	1000–2500
Shear strength MPa	40		10		25
Compressive strength MPa	100		25		50
Impact strength J cm^{-1}	0.5 notched		0.4 notched		0.2–0.5 notched
Flexure strength MPa	150		80		10–100
Strain at failure %	5	(29)	500–1000	(30)	2.5–10
Thermal expansion coefficient K^{-1}	3×10^{-5}		1×10^{-4}		5×10^{-4}
Thermal conductivity J m^{-1}s^{-1}K^{-1}	0.5		0.2		0.5
Specific heat KJ Kg^{-1}K^{-1}	1.4		1.5		1.5

	DA Complex of Polyesters and Polycarbonates		TTF:TCNQ Containing Polyurethane	Metal Ion Coordination Polymers
Glass transition temperature K	363	(29)	250	dec.
Dielectric constant	3.2		3.0	1.0
Electrical resistivity Ωcm	10^{12}-10^{13}		10^{5}-10^{7}	10^{-1}-10^{13}
Dielectric strength	500		250	300
Surface resistivity Ω [per square]	10^{12}		10^{5}	10^{-1}
Dissipation Factor	0.003		0.005	0.001
Temperature range K	100-450		100-400	100-600
Outgassing	TML 0.2 CVCM 0.05		TML 0.5 CVCM 0.05	TML 0.1 CVCM 0.05
Ionizing radiation MRad	5-50		1-10	10-50
Heat distortion temperature K	400		250	250-500

	Polyacetylene (Doped)		Polydiacetylenes		Polyphenylacetylene	
Specific gravity $Kg\ m^{-3}$	100–400	(19)	1370	(18)	1200	
Tensile strength MPa	3		2–17	(18)	10	
Tensile modulus MPa	10		42	(18)	25	
Shear strength MPa	5		25		20	
Compressive strength MPa	5		10		15	
Impact strength $J\ cm^{-1}$	0.05 notched		0.3 notched		0.3 notched	
Flexure strength MPa	75		200		150	
Strain at failure %	⩽300	(20)	⩽3	(18)	10	
Thermal expansion coefficient K^{-1}	5×10^{-5}		5×10^{-5}		5×10^{-5}	
Thermal conductivity $J\ m^{-1}s^{-1}K^{-1}$	$2\text{-}6 \times 10^{-3}$	(21)	5×10^{-3}	(21)	2.30×10^{-3}	(21)
Specific heat $KJ\ Kg^{-1}K^{-1}$	1.4		1.4		1.2	

	Polyacetylene (Doped)	Polydiacetylenes	Polyphenylacetylene
Glass transition temperature K	decomposes (20)	decomposes (18)	350
Dielectric constant	3	3	3
Electrical resistivity Ωcm	10^{-3}-10^{12}	10^{15}	10^4-10^{12}
Dielectric strength	500	1000	1000
Surface resistivity Ω [per square]	10^{-3}	10^{15}	10^4
Dissipation Factor	0.001	0.001	0.003
Temperature range K	100-350	100-400	100-400
Outgassing	TML 0.1 CVCM 0.03	TML 0.1 CVCM 0.03	TML 0.5 CVCM 0.03
Ionizing radiation MRad	1	1	10
Heat distortion temperature K	dec.	350	dec.

REFERENCES FOR APPENDIX E

1.   Product Assurance Division European Space Research and Technology
     Center. Guidelines for Space Materials Selection, ESA PSS-07 (QRM-01)
     Issue 5 (July 1979).

2.   M. Naitove, editor, Plastics Technology, 26(6) (mid-May 1980).

3.   Plastics Desk Top Data Bank, Book B, fifth edition, M. Howard, editor
     (International Plastics Selector, Inc., San Diego, CA, 1980).

4.   J. Agranoff, editor, Modern Plastics Encyclopedia 59(10A) (1982).

5.   Polymer Handbook, 2nd Edition, J. Bendrup & E. Immergut Eds., (Wiley
     Interscience, NY, 1975).

6.   A Ready Reference for Plastics (Boontan Moulding Company, 1973).

7.   J. Rittenhouse and J. Singletary, Space Materials Hand Book, Third
     edition, AFML-TR-68-205 (1966).

8.   LNP Engineering Plastics Fortified Polymers. Bulletin 202-877 (1977).

9.   D. A. Seanor., J. of Poly. Sci.: Part A-2, 6, 463 (1968).

10. N. R. Lerner, J. Appl. Phys. 52(11), (1981).

11. E. D. Goodings, Chem. Soc. Reviews 5, 95 (1976).

12. T. C. Clarke et al., IBM Research Report RJ2945 (36890) (September 22,
     1980).

13. John A. Mock, Handbook of plastics and Elastomers (Morgan Hill, NY, 1975.

14. A. Fischer and B. Mermelsteint, NASA TMX-65705 (1971).

15. J. Rittenhouse et al., Space Materials Handbook Supplement, 1 to the 2nd
     Edition, AFML-TR-62-185 (1962).

16. K. Keiji Kanazawa et al., Synth. Metals 1, 329 (1979/80)

17. J. H. Lupinski and K. D. Kopple, Science 44, 1038, (1964).

18. R. H. Baughman et al., J. of Polym. Sci.: Poly. Phys. Ed. 13, 1871
     (1975).

19. G. E. Wnek, J.C.W. Chien, F. E. Karasz, M. A. Druy, Y. W. Park, A. G.
     MacDiarmid, and A. J. Heeger, Polym. Prepr. 21, 447 (1982).

20. Y. W. Park et al., J. Poly. Sci.: Polym. Lett 17, 195 (1979)

21. R. K. Jenkins and N. R. Byrd, Final Report NASA CR-193573 (1968).

22. C. Hsu et al., J. Chem. Phys. 61, 4640 (1974).

23. Stephen D. Bruck, J. of Polym. Sci.: Part C 17, 169 (1967).

24. M. Jozefowicz et al., J. of Polym. Sci.: Part C 22, 1187 (1969).

25. S. R. Allen et al., Macromolecules 14, 1135, (1981).

26. R. Bake and D. Chen in Proceedings of the Conference on Electrical Insulation and Dielectric (National Research Council, Boston, MA, 1981, p. 297).

27. C. J. Norrell et al., J. of Polym. Sci.: Polym. Phys. Ed. 12, 913 (1974).

28. K. Saha et al., J. of Noncrystalline Solids 22, 291 (1976).

29. T. Sulzberg and R. J. Cotter, J. of Polym. Sci.: Part A-1 8, 2747 (1970).

30. M. Watanabe et al., J. of Poly. Sci.: Polym. Lett. Ed. 19, 331 (1981).

Appendix F
Compilation of Polymer Conductivity Data

Electrical resistivity data for a range of commercial and experimental polymers are summarized in Appendix F along with comments regarding the experimental conditions under which measurements were made and the dependence of conductivity on temperature or radiation. Materials are organized into general classes designated by a Roman numeral, the specific literature reference by an arabic number, and multiple materials in a given reference by lowercase letters. For example, V6a is one specific polymer structure found in reference V6, which is grouped with polyacetylene derivatives in Section V.

Polymer / Reference	Resistivity, Ω cm	Experimental Variables	Functional Dependence	Comments
I Polyimide				
1	$10^{-1} - 10^{3}$	Pyrolysis at 650°-850°C	-	50 MHz, RT
2	$10^{-2} - 18^{18}$	Pyrolysis	-	Commercial and "ultra-pure" materials
3	$2.5 \times 10^{9} - 10^{14}$ $2 \times 10^{7} - 4 \times 10^{9}$	Temperature Temperature	- -	DC 1-ns transients
4	$10^{18} - 10^{15}$	Pyrolysis	-	RT
5	$10^{13} - 10^{15}$	Time, electric field, heat vacuum hν electron-beam	-	
6	-	Electron-beam	-	20 kV, 80 nA/cm²
7	6×10^{14} 4×10^{16}	Beam energy	-	2.5-20 keV electron-beam discharges begin at 12 keV.
8	$10^{10} - 10^{13}$	Beam energy	-	Metalized Kapton electron beam 0.2-10 keV charging reduced at beam energies greater than 6 keV.
9	-	Electron-beam	-	Surface charging of composite.

Polymer / Reference	Resistivity, Ω cm	Experimental Variables	Functional Dependence	Comments
10	—	—	—	Composite structures of carbon, carbon fiber, adhesive, and Kapton. Potential charge decreases with increasing carbon content.
II Teflon				
1	$10^{13} - 10^{17}$	Temperature, field dose rate	$I\alpha R^{\Delta} - \Delta = 0.63$	—
2	8×10^{17}	—	—	20°C, 8 rad/min. E_a 0.5 eV
3	3.26×10^{13} - 1.25×10^{15} radiation induced	—	$\sigma_o = 6.4 \times 10^{-8}|I_o|^{0.79}$ I_o $10^{-8} - 10^{-10}A$	40 Kev electron beam.
4	4.5×10^{14} - 1×10^{15}	Dose rate, electric field.	$\sigma_{ss} = 7.5 \times 10^{-17}$ $\times (R/S)^{0.64}$ $\sigma_{ss} = \sigma_o (E)^{0.35}$	20-220 rad/sec.
5	—	Electron beam, electric field	—	6.30 keV electron-beam
6	$\sim 10^{13}$	Hardening, pressure	—	Radiation hardened Teflon
7	—	Heat, hν	—	—
8	$2 \times 10^{17} - 5 \times 10^{20}$ initial 10^{14} on γ irradiation	Radiation	$I\alpha$ dose rate	2 MeV γ and Co^{60}

Polymer Reference	Resistivity, Ω cm	Experimental Variables	Functional Dependence	Comments
9		Dose rate temperature	$\sigma \propto R^{1/\alpha_T - (1-1/\alpha)} \exp[-(1-1/\alpha)Q/kT]$	
III Polyethylene Terephthalate				
1	$10^{18} - 10^{21}$	Dose rate temperature	$\sigma = \sigma_0 \exp(-1.3 \text{ eV}/kT)$ $I\alpha$ dose $^{0.83}$	E_a decreases to 0.2 eV at 7 rads/min.
2	—	Temperature	$\mu = \mu_0 \exp(-E_a/kT)$ $E_a = 0.2 - 0.3$ eV	Negatively charged carriers
3	$6.4 \times 10^9 - 9 \times 10^{14}$	Electric field, temperature	$I\alpha V^{112}$	2-1400 kV/cm, 75-150°C
4	—	Radiation dose rate, thickness	$I \alpha R^\Delta$ $0.5 < \Delta < 1.0$	Δ thickness dependent
5	5×10^{12}	Electric field	—	R oscillates at high (MV/cm) fields.
6	$10^{14} - 10^{18}$	Temperature	$\sigma = 1.2 \times 10^7 \exp(-1.7 \text{eV}/kT)$ $\sigma = 8 \times 10^{11} \exp(-2.2 \text{ eV}/kT)$	—
IV Polyethylene/Polypropylene				
1	$10^{15} - 10^{20}$	Radiation dose rate	$I\alpha R^\Delta$, $\Delta = 0.8$ $\sigma = \sigma_0 \exp(-E_a/RT)$	γ exposure
2	$1 \times 10^{12} - 5 \times 10^{12}$	Dose rate	$E_a = 0.35$ eV $\sigma_0 = 9 \times 10^{-17}$	200 - 1200 rads/min

Polymer / Reference	Resistivity, Ω cm	Experimental Variables	Functional Dependence	Comments
3	$2 \times 10^{14} - 1 \times 10^{16}$	Dose rate	—	$10^3 - 10^5$ rad/hr
4	—	Dopant	—	Mobility increases by 10^4 with iodine doping.
V Polyacetylene and Derivatives				
1	$10^{-3} - 10^9$	Dopant	—	Cis isomer
2	$4 \times 10^{-3} - 3 \times 10^{-4}$	Orientation	—	—
3	$10^5 - 10^{12}$	Temperature	—	Trans isomer
4	$10^4 - 10^{12}$	Dopant crystallinity	$\sigma = \sigma_0 \exp^{-E_a/RT}$ $E_a = 0.2$ eV $\sigma = \sigma_{max}[\gamma \cdot C(1-C)]$ $C \equiv (\gamma/4)$ mole fraction iodine $\sigma_{max} = 0.4 \times 10^{-5}$ $\sigma_{max} = 1.8 \times 10^{-5}$	Amorphous & crystalline materials; Amorphous $\gamma = 96$; Crystalline $\gamma = 205$
5	—	Orientation hν	$\sigma_{11}/\sigma_{\perp} = 6$ (dark) $\mu_{11}/\mu_{\perp} = 800 \pm 300$ (light)	
6a	5×10^{18}	—	$\sigma_0 = 1 \times 10^{-3}$ $E_a = 0.85$ eV	Thermal polymerization

Polymer / Reference	Resistivity, Ω cm	Experimental Variables	Functional Dependence σ_o	E_a (eV)	Comments
6					
b	7.7×10^{17}	–	5.5×10^{-3}	0.91	Thermal polymerization
c	7.7×10^{17}	–	1.4×10^{-3}	0.87	Thermal polymerization
d	1.4×10^{17}	–	1.9×10^{8}	1.48	–
e	$>10^{20}$	–	–	–	–
f	2.3×10^{17}	Copolymer	6.9×10^{-3}	0.88	–
g	2.3×10^{17}	Copolymer	3.6×10^{-2}	0.92	–
h	$>10^{20}$	Copolymer	–	–	–
i	3.4×10^{17}	Dopant	1.1×10^{12}	1.75	CCl_3COOH
j	2.4×10^{13}	Dopant	4.2×10^{-5}	0.53	CCl_3COOH
k	2.2×10^{8}	Dopant	3.2×10^{2}	0.63	CCl_3COOH
l	1×10^{15}	Dopant	3.8×10^{-4}	0.68	Tetracyanoethylene
m	3×10^{15}	Dopant	1.2×10^{-2}	0.74	Tetracyanoethylene
n	1.3×10^{15}	Dopant	1.9×10^{8}	1.41	N vinyl carbazole
o	4.0×10^{17}	Dopant	7.4×10^{-4}	0.84	N vinyl carbazole
7	$3 \times 10^{0} - 1 \times 10^{12}$	Dopant	–	–	ASF_5

Polymer Reference	Resistivity, Ω cm	Experimental Variables	Functional Dependence	Comments
8a	8×10^{15}	–	–	–
b	$1 \times 10^{10} - 5 \times 10^{13}$	Dopant	–	TCNQ
c	8.9×10^{6}	Dopant	–	I_2
d	1.2×10^{15}	–	–	–
e	1.3×10^{11}	Dopant	–	TCNQ
9a	1.5×10^{11}	–	$E_a = 0.65$ eV	–
b	3×10^{9}	–	–	–
10	10^{12}	–	–	–
11	10^{6}	–	$E_a = 0.19$ eV	–
12	$10^{10} - 10^{13}$	–	–	–
13	10^{8}	–	$E_a = 0.42$ eV	–
VI Polysulfur Nitride				
1	5×10^{4} 1×10^{-1}; $10^{3} - 10^{14}$	Preparation temperature	–	50-500 fold change due to orientation with respect to chain axis.
2	50	–	–	10-15 fold change due to orientation expitaxial growth on Mylar.

Polymer Reference	Resistivity, Ω cm	Experimental Variables	Functional Dependence	Comments
3	2×10^{-7} 2×10^{-4}	@ 0.07°K @ 1°K	–	–
VII Polyacylene Quinone Radical Polymers				
1a	$10^2 - 10^5$ $10^1 - 10^5$ $10^1 - 10^5$	Frequency DC Temperature	–	–
b	$10^5 - 10^9$ 10^5 $10^5 - 10^{11}$	Frequency DC Temperature	–	–
2a	1.72×10^6		E_a (eV) = 0.21	300 K, 1820 bars
b	8.65×10^7		0.34	300 K, 1820 bars
c	5.78×10^5		0.25	300 K, 1820 bars
d	1.6×10^6		0.24	300 K, 1820 bars
e	4.25×10^5		0.25	300 K, 1820 bars
f	4.2×10^4		0.20	300 K, 1820 bars
g	2.2×10^5		0.30	300 K, 1820 bars
h	6.65×10^4		0.31	300 K, 1820 bars
i	1.08×10^4		0.26	300 K, 1820 bars
j	1×10^3		0.20	300 K, 1820 bars

Polymer Reference	Resistivity, Ω cm	Experimental Variables	Functional Dependence	Comments
3a	9×10^4		E_a (eV) = 0.18	25°C/6.3 k bar
b	4×10^6		0.225	25°C/6.3 k bar
c	5.3×10^6		0.24	25°C/6.3 k bar
d	9×10^5		0.28	25°C/6.3 k bar
e	6×10^7		0.70	25°C/6.3 k bar
4a	$2.5 - 6 \times 10^{11}$			
b	6×10^{11}			
c	5×10^{11}			
VIII Vinyl-Carbazole-Containing Polymers				
1	$10^9 - 10^{16}$	Copolymer composition	E_a = 0.12 - 1.3 eV	
2	$5 \times 10^{10} - 1 \times 10^5$	Dopant	E_a = 0.39 - 1.27 eV	Iodine
3a	10^{14}	Dopant		Radical cation
b	$10^5 - 10^{13}$	Dopant		
IX Polyvinyl Pyridine				
1a	10^{15}		–	–
b	$10^4 - 10^7$	Dopant	–	Iodine

Polymer Reference	Resistivity Ω cm	Experimental Variables	Functional Dependence E_a (eV)	ρ_o	Comments
2a	0.6×10^9		0.40	1.67×10^3	Atactic TCNQ complex
b	4.44×10^8		0.505	2.97×10^{-1}	Atactic TCNQ complex
c	8.39×10^6		0.65	1.19×10^{-4}	Atactic TCNQ complex
d	1.4×10^{13}		1.22	7.16×10^{-6}	Atactic TCNQ complex
e	4.68×10^3		0.11	6.46×10^1	Isotactic TCNQ complex
f	2.76×10^4		0.14	0.76×10^1	Isotactic TCNQ complex
g	1.46×10^5		0.18	5.18×10^1	Isotactic TCNQ complex
h	6.05×10^4		0.20	3.03×10^1	Isotactic TCNQ complex
3a	$10^4 - 10^6$	Dopant	–	–	TCNQ complex
b	$10^4 - 10^9$	Dopant	–	–	Polystyrene copolymer
X Phthalocyanine Polymers					
1	$3 \times 10^7 - 1 \times 10^{10}$	Composition	E_a (kcal/mol) = 6.5 – 7.5		
2	6.8 – 5.8	–	–		Metal free
3	$10^8 - 10^{12}$	Temperature	–		Monomeric
4	3 – 100	–	–		Coordinated

Polymer Reference	Resistivity, Ω cm	Experimental Variables	Functional Dependence	Comments
5	$3 \times 10^1 - 1 \times 10^8$	Dopant	–	
6	$4 \times 10^3 - 1.4 \times 10^3$	Composition	$\sigma_o = 1 \times 10^{-10} - 8 \times 10^{-3}$ E_a (eV) $= 0.02 - 1.13$	Variations in metal chelate and organic report units.
7	$1.6 \times 10^1 - 1.6 \times 10^2$ $10^{-1} - 10^{12}$ $10^3 - 10^{12}$	Temperature	–	Metal free
8	5×10^{15} 1.7×10^6	Extent of cure	E_a $0.6 - 2.1$ eV $\sigma_o = 3 \times 10^{-10}$ to 2.7×10^6	–
XI TTF: TCNQ Containing Polymers				
1a	5×10^7	–	–	Neutral salt complex
b	1×10^6	–	–	Charged salt complex
2a	3×10^2	–	–	Ferrocene
b	6×10^6	–	–	TTF
c	5×10^5	–	–	TTF:I_2
3	$5 \times 10^7 - 5 \times 10^8$	Elongation	–	–
4	–	–	–	Unable to form complex

Polymer / Reference	Resistivity, Ω cm	Experimental Variables	Functional Dependence	Comments
5	$10^{14} - 2 \times 10^8$	-	-	-
6	$>10^6 - 10^{16}$	Temperature, molecular weight	-	Elastomers
7	$10^3 - >10^{10}$	-	-	TCNQ: Poly cation complex
XII Metal Ion Containing Polymers				
1a	$3 \times 10^6 - 1.25 \times 10^{11}$	Composition		
b	$2.5 \times 10^9 - 1.5 \times 10^{13}$	Oxidation state		
c	$6 \times 10^5 - 2 \times 10^{11}$	Oxidation state		
2	2.5×10^{-1}	-	-	Mixed valence
3a	10^6	Oxidation state	-	Fe II
b	$10^7 - 10^8$	Oxidation state	-	Fe II, III
c	10^6	Oxidation state, dopant		I_2, Fe II, III
4	$2.5 \times 10^4 - 1 \times 10^6$	Temperature		Cl, Ni, Pd complexes
5	$10^{-12} - 10^2$ $10^3 - 10^5$	Orientation, temperature	-	-

Polymer Reference	Resistivity, Ω cm	Experimental Variables	Functional Dependence	Comments
6a	6.3×10^{15}	Dopant	E_a (kcal/mol) >15	–
b	1.25×10^7	Dopant	16	Iodine
c	$\sim 1 \times 10^{10}$	Dopant	$11.1 - 16$	Tetracyanoethylene
7	$4 \times 10^4 - 1 \times 10^{10}$	Crystallinity	$E_a = 0.45$ (crystalline) 0.22 (amorphous)	
XIII Polyamides				
1	$10^{11} - 10^{14}$	Composition	E_a (eV) $= 1.3-1.5$	Various proteins
2	$2 \times 10^8 - 2.6 \times 10^{10}$	Composition	E_a (eV) $= 0.5-0.6$	Aliphatic polymers
3	–	Temperature range	E_a (eV) $= 1.3-2.2\,eV$ $\sigma_o = 5 \times 10^6 - 3 \times 10^{17}$	
4	$10^{11} - 10^{12}$	Radiation dose	–	200 keV x-ray Co^{60} γ
5	–	–	$\sigma_o = 7 \times 10^3$ $E_a = 2.2$ eV (dry) 2.7 eV (wet)	
6		Temperature, orientation	$2 \times 10^2 < \sigma_o < 3 \times 10^{20}$ $1.02 < [E_a(eV)] < 2.28$	
7	–	–	–	Crystallization above the glass transition temperature increases σ by 10 - 100 times

Polymer Reference	Resistivity, Ω cm %RH	Experimental Variables ρ_{11}	ρ_{22}	Functional Dependence E_{11} (kcal/mol)	E_{22}	Comments
XIV Other Polymers						
1,2	0	10^{12}	2×10^{17}	9.9	20.7	Ionic conductivity indicated
	40	8×10^{10}	2.6×10^{16}	8.8	-	
3	1.6×10^4					
4a	1.3×10^{16}	Composition		-		Polymeric donor–acceptor complexes
b	1×10^{13}					
c	1×10^{13}					
d	1×10^{13}					
e	2.7×10^{13}					
f	1×10^{12}					
g	2.5×10^{11}					
5	Electric field and radiation dose rate	$I\alpha$ dose rate Δ 0.5 dark		900 rad/min		
a	$\sim 10^{16}$	$10^{18} - 5 \times 10^{19}$	$10^{17} - 5 \times 10^{18}$	$\sim 10^{14}$		
b						

Polymer Reference	Resistivity, Ω cm	Experimental Variables	E$_a$ (eV)	σ$_o$	Comments
6	$10^{-1} - 10^4$	Temperature, hydration	0.3–2.0	$10^{-7} - 10^6$	
7	$10^4 - 10^{20}$	Composition	0.35–1.08	$10^{-8}-10^{-3}$	
	$5 \times 10^5 - 5 \times 10^{14}$	Chain length			
8a	$10^{15} - 10^{18}$	Dopant, temperature	–		Undoped
b	$10^0 - 10^3$		–		AsF$_5$ dope
9a	10^{17}	Dopant	–		Undoped
	4.2×10^7	Dopant concentration	0.69		
	$1 \times 10^8 - 1 \times 10^{16}$		0.44 – 1.11		TCNQ
b	10^{17}		–		Undoped
	$2.5 \times 10^9 -$ 6×10^{15}	Dopant concentration	0.5 – 0.76		p-fluoranile
	$2.4 \times 10^7 -$ 2.2×10^9	Dopant concentration	0.43 – 0.55		TCNQ
10a	2.2×10^{12}		0.45–0.6	2.1×10^{-5}	Linear
b	4.5×10^{12}		0.35–0.66	7×10^{-5}	Branched
c	2.1×10^{12}		0.63	3×10^{-5}	Branched
d	1.9×10^{11}		0.17–0.56	8.9×10^{-5}	Linear
e	7.0×10^{11}		0.17–0.57	3.1×10^{-5}	Linear
f	2×10^{12}		0.57	5.0×10^{-7}	Linear

Polymer Reference	Resistivity, Ω cm	Experimental Variables	Functional Dependence		Comments
g	3.6×10^{12}		1.4×10^{-6}	0.52	Linear
h	2.3×10^{12}		1.4×10^{-7}	0.35	Branched
11					$\mu + 10^{-5}$ to 10^{-8} (cm^2/V sec)
12a	2.3×10^{4}	Neutral			
	7.7×10^{6}	Reduced			
b	5×10^{8}	Neutral			
	1.8×10^{9}	Reduced			
13a	3.8×10^{11}	Composition			
b	5×10^{9}				
c	1.25×10^{7}				
d	2×10^{7}				
14a	1.08×10^{6}	Composition			
b	5.8×10^{5}				
c	1.25×10^{7}				
d	6.30×10^{5}				
e	5×10^{7}				
15	$6 \times 10^{7} - 2 \times 10^{8}$				

Polymer Reference	Resistivity, Ω cm	Experimental Variables	Functional Dependence	Comments
16	1.2×10^9			
17	$10^1 - 10^6$	Composition		
18	10^{-1}	Various vinyl acceptors		
19a	$10^3 - 10^{10}$	Temperature		
b	$10^3 - 10^{10}$	Composition		
c	5×10^2			
d	2×10^3			
20	$10^7 - 10^{15}$	Temperature, pressure		
21	$<10^{15}$	–		
22	$5 \times 10^8 - 10^{16}$	Composition		
23	$10^1 - 10^{-2}$	pH, hydration composition		
24	$2 \times 10^2 - 1 \times 10^{-2}$	Synthesis method		
25	8×10^{10}			
26a	$4.6 \times 10^{12} - 1 \times 10^{15}$	Polymerization method	$E_a = 1.32-1.67$ eV	
	$2.1 \times 10^6 - 3.2 \times 10^{10}$	Dopant concentration	$E_a = 1.01-1.36$ eV	Iodine
b	10^{16}		$E_a = 1.49$ eV	

Polymer Reference	Resistivity, Ω cm	Experimental Variables	Functional Dependence E_a (eV)	Functional Dependence σ_o	Comments
27a	1.4×10^{13}	–	–		–
	$\sim10^{15}$	Dopant	–		TCNE
	$\sim10^{14}$	Dopant	–		Chloranil
	10^{13}	Dopant	–		DDQ
b	1.5×10^{13}	Dopant	–		TCNE
	1.8×10^{13}	–	–		–
28a	5×10^{12}		2.23	7.2×10^{-7}	25°C, 1800 atm
b	9×10^{11}		2.15	5.6×10^{-7}	25°C, 1800 atm
c	1.8×10^{12}		1.57	9.5×10^{-2}	25°C, 1800 atm
d	1.4×10^{13}		2.15	8.7×10^{-6}	25°C, 1800 atm
e	3.2×10^{15}		1.66	2.9×10^{1}	25°C, 1800 atm
f	7.5×10^{14}		1.91	5.4×10^{-2}	25°C, 1800 atm
g	1×10^{14}		1.78	8.1×10^{-2}	25°C, 1800 atm
h	4×10^{10}		1.12	1.35×10^{1}	25°C, 1800 atm
i	5.7×10^{13}		1.56	3.6×10^{0}	25°C, 1800 atm
j	1.1×10^{16}		2.9	3.2×10^{-9}	25°C, 1800 atm
29					

Polymer / Reference	Resistivity, Ω cm	Experimental Variables	E_a (eV)	σ_o	Comments
30					
31a	2.5×10^7 – 1.4×10^{15}	Metal ligand	$0.6 - 0.7$	10^{-3}–10^4	
b	1.4×10^{10} – 2×10^{12}	Composition	0.36–0.72	10^{-3}–10^0	
c	5.9×10^{11} – 1.2×10^{12}	Composition	0.58–0.76	1×10^{-3} / 5×10^{-3}	
32	10^4 – 10^9	–	–		–
33a	10^2	Heat treatment, composition	E_a (eV) = 1.1 – 1.5		
b	$\sim 10^2$	Composition	1.5		
c	$\sim 10^2$	Composition	0.75		
d	$\sim 10^1$		0.66		
34a	10^7 – 10^{13}	Oxidation state	1.0		
b	10^{11} – 10^{14}		1.25		Linear crosslinked materials
c	10^6 – 10^{10}		0.8		
d	10^6 – 10^{10}		1.0		

Polymer Reference	Resistivity, Ω cm	Experimental Variables	Functional Dependence E_a (eV)	Functional Dependence ρ_o	Comments
35a	5×10^{11}	Composition	1.8	8×10^{0}	
b	1.3×10^{12}	Composition	1.5	1.8×10^{-2}	
c	6.6×10^{13}	Composition	1.8	1.2×10^{-2}	
d	1×10^{14}	Composition	2.6	5×10^{-3}	
e	1.8×10^{14}	Composition	2.5	8×10^{-2}	
f	2.5×10^{13}	Composition	1.8	6×10^{-2}	
g	1.6×10^{14}	Composition	2.2	6×10^{0}	
h	2.2×10^{14}	Composition	2.2	1.2×10^{0}	
i	1×10^{15}	Composition	2.6	3×10^{-2}	
36a	5×10^{13}		0.75	6.9×10^{-8}	
b	3.3×10^{14}		0.70	5.3×10^{-9}	
c	4×10^{14}		0.87	8.0×10^{-8}	
d	2.5×10^{11}		0.87	9.1×10^{-5}	
e	1.4×10^{15}		0.60	8.5×10^{-11}	
f	3×10^{13}		0.88	1.1×10^{-6}	

Polymer / Reference	Resistivity, Ω cm	Experimental Variables	Functional Dependence	Comments
g	6×10^{10}		0.89 3.0×10^{-3}	
h	1×10^{15}		0.71 1.0×10^{-8}	
i	1.3×10^{15}		0.89 2.3×10^{-5}	
j	$\sim 10^{16}$		0.78 5.5×10^{-10}	
k	$\sim 10^{16}$		– –	
l	2.5×10^{15}		– –	

REFERENCES FOR APPENDIX F

I   Polyimides (Kapton)

1. S. D. Bruck, Polymer 6, 319 (1965).

2. H. Brom et al., Solid State Communications 35(2), 135 (1980).

3. J. Hanscomb and J. Calderwood, J. Phys. D 6(9), 1093 (1973).

4. J. Lin et al., ORPL 43(2), 482 (1980).

5. H. Coffey et al., NASA CR134995 (1976).

6. K. Balmain and G. Dubois, IEEE-NS 26(6), 5146 (1979).

7. J. Staskus and F. Berkopec, "Test Results for Electron Beam Charging of Flexible Insulators and Composities," in " Spacecraft Charging Technology," R. Finke and C. Pilce, editors, AFGL-TR-79-0082 (1979), p. 457.

8. P. Robinson, Jr., J. Spacecraft and Rockets 16(2), 104 (1979).

9. C. Fellas, IEEE-NS 27(6), 1801 (1980).

10. J. Staskus and S. Jarciso, NASA TM-73865 (1978).

II   Teflon

1. A. Lilly and J. McDowell, J. Appl. Phys. 39(1), 141 (1968).

2. J. F. Fowler, Proc. Roy. Soc. London A236, 464 (1956).

3. B. Gross et al., J. Appl. Phys. 51(9), 4875 (1980).

4. Ibid., 52(2), 571 (1981).

5. J. Jog et al., Polymer 22(7), 865 (1981).

6. B. Gross et al., Appl. Phys. Lett. 24(8), 351 (1974).

7. R. Adamo and J. Nanevicz, NASA CR 135201 (1977).

8. R. Meyer et al., J. Appl. Phys. 27(9), 1012 (1956).

9. K. Yahagi and A. Danno, J. Appl. Phys. 34(4), 805 (1963).

III   Polyethylene Terephthalate (Mylar)

1.   J. Folwer and F. Farmer, Nature 175, 590 (1955).

2.   E. Martin and J. Hirsch, J. Appl. Phys. 43(3), 1001 (1972).

3.   A. Lilly and J. McDowell, J. Appl. Phys. 39(1), 141 (1968).

4.   H. Maeda et al., J. Appl. Phys. 52(2), 758 (1979).

5.   P. Predecki and N. Swaroop, J. Polym. Sci. Lett. Ed. 9(1), 43
      (1971).

6.   E. Sacher, J. Macro. Sci. Phys. B4(2), 441 (1970).

VI   Polyethylene, Polypropylene

1.   J. Fowler, Proc. Roy. Soc. London A236, 464 (1956).

2.   R. Meyer et al., J. Appl. Phys. 27(9), 1012 (1956).

3.   K. Yahagi and A. Danno, J. Appl. Phys. 34(4), 804 (1963).

4.   D. Davies, J. Phys. D 5(1), 162 (1972).

V   Polyacetylene and Derivatives

1.   Y. Park et al., J. Chem. Phys. 73(2), 946 (1980).

2.   Y. Park et al., J. Polym. Sci., Lett. Ed. 17, 195 (1979).

3.   Y. Yamamoto et al., Jap. J. Appl. Phys. 19(5), 991 (1980).

4.   P. Cukor et al., Makro Chemie 182(1), 165 (1981).

5.   K. Lochner et al., Chem. Phys. Lett. 41(2), 388 (1976).

6.   A. Hankin and A. North, Trans Far. Soc. 63, 1525 (1967).

7.   G. Wenk et al., Polymer 20(12), 1441 (1979).

8.   R. Jenkins and N. Byrd, NASA CR 103573 (1968).

9.   H. Pohl and R. Chartoff, J. Polym. Sci. A2, 2787 (1964).

10.   S. Kambra et al., J. Polym. Sci. B5, 233 (1967).

11.   G. deVries and L. Van Beek, Rec. Trav. Chim. 84, 184 (1965).

12. J. Stille and D. Frey, JACS 83, 1697 (1961).

13. I. Storbeek and M. Starke, Z. Electrochem. 69, 343 (1965).

VI   Polysulfur Nitride

1.   C. Hsu et al., J. Chem. Phys. 61, 4640 (1974).

2.   A. Bright, Appl. Phys. Lett. 26, 612 (1975).

3.   R. Green and B. Street, Phys. Rev. Lett. 34, 577 (1975).

VII Polyacylene Quinone Radical Polymers

1.   H. Pohl et al., J. Non-Crystalline Solids 22, 291 (1976); ibid. 21, 117 (1976).

2.   H. Pohl et al., J. Polym. Sci. Symposium 17, 191 (1967); also see H. Pohl and J. Wyhof, J. Polym. Sci. A1 10(2) 387 (1972).

3.   H. Pohl and R. Hartman, J. Polym. Sci. A1, 6(5), 1135 (1968).

4.   H. Pohl and D. Opp, J. Phys. Chem. 66, 2085 (1962); also see H. Pohl and E. Engelhardt, J. Phys. Chem. 66, 2121 (1962).

VIII   Vinyl-Carbazole-Containing Polymers

1.   C. Pittman and P. Grube, J. Appl. Polym. Sci. 18(8), 2269 (1974).

2.   A. Herman and A. Rembaum, J. Poly. Sci. Symposia 17, 107 (1967).

3.   H. Block et al., Polymer 18(8), 781 (1971).

IX   Polyvinyl Pyridine

1.   S. Mainthia et al., J. Chem. Phys. 41, 2206 (1964).

2.   D. Bonnifore et al., Disc. Farady Soc. 51, 131 (1971).

3.   J. Lupinski and K. Kopple, Science 142, 1039 (1964).

X   Phthalocyanine Polymers

1.   J. McKellar et al., Disc. Farady Soc. 51, 176 (1971).

2.   S. Burnay and H. Pohl, J. Non-Crystalline Solids 30, 221 (1978).

3.   H. Nalwa et al., Makro Chemie 182, 811 (1981).

4.   A. Epstein and B. Wildi, J. Chem. Phys. 32(2), 324 (1960).

5.   T. Kellar and J. Griffith, NRL Memorandum Rept. 4345 116 (1980).

6.   G. Manecke and D. Wohrle, Makro Chemie 120, 176 (1968).

7.   T. Walton and J. Reardon, NRL Memorandum Rept. 3960 80, 98 (1979).

8.   G. Meyer and D. Wohrle, Makro Chemie 175, 714 (1974); also see S.
     Shrivastava and A. Stivastava, Polymer 22(6), 765 (1981).

XI   TF: TCNQ-Containing Polymers

1.   M. Watanabe et al., J. Polym. Sci. Lett. Ed. 19, 331 (1981).

2.   W. Hertler, J. Org. Chem. 41(8), 1412 (1976).

3.   A. Hermann et al., J. Polym. Sci., Lett. Ed. B9(8), 627 (1971).

4.   C. Pittman et al., Macromolecules 12 (3,4), 355, 541 (1979).

5.   J. Summers and M. Litt, J. Polym. Sci. Chem. 11, 1379 (1973).

6.   A. Rembraum et al., J. Polym. Sci., Lett. Ed. 8, 467 (1970).

7.   J. Lupinski et al., J. Polym. Sci. Symposia 16, 1561 (1967).

XII   Metal-Ion-Containing Polymers

1.   D. Cowan et al., JACS 94, 5110 (1972).

2.   M. Minot and J. Perlstein, Phys. Rev. Lett. 26(7), 371 (1968).

3.   C. Pittman et al., JACS 96(26), 7916 (1974).

4.   M. Dewar and A. Talati, JACS 86, 1592 (1964).

5.   H. Zeller, J. Phys. Chem. Solids 35, 77 (1974).

6.   J. McKellar et al., Disc. Farady Soc. 51, 176 (1971).

7.   S. Kanda et al., J. Polym. Sci. Symposia 17, 151 (1967); also see
     F. Gotzfried el al., Angew Chem. 18(6), 463 (1979).

XIII Polyamides

1. D. Eley and D. Spivey, Trans. Farady Soc. 56, 1432 (1960).

2. Ibid. 57, 2280 (1961).

3. D. Senor, J. Polym. Sci. A2 6, 463 (1968).

4. P. Itedvig, J. Polym. Sci., Chem. Ed. 2, 4097 (1964).

5. B. Rosenberg, J. Chem. Phys. 36(3), 816 (1962).

6. D. Seanor, J. Polym. Sci. Symposia 17, 195 (1967).

7. B. Sazhin anbd N. Podesenova, Vysokomol. Soedin 6, 137 (1964).

XIV Other Polymers

1. R. Baker et al., submitted to Conf. Elect. Ins. Diel. Phenomena (1981).

2. R. Baker and K. Lawless. Annual Report UVA 525630 (1981).

3. P. Kovacic et al., J. Polym. Sci., Lett. Ed. 19, 347 (1981).

4. T. Sulzberg and R. Cotter, J. Polym. Sci. A1 8, 2747 (1970).

5. P. Rencroft, J. Appl. Polym. Sci. 14, 1361 (1970).

6. M. Jozefowicz et al., J. Polym. Sci. 14, 1361 (1970).

7. G. Kossmehl et al., Makro Chemie 29, 307 (1973); see also Angew Chemie 29/30 307 (1969).

8. T. C. Clark et al., IBM RJ 2945 (1980).

9. R. Knoesel et al., Bull. Chem. Soc. Fr. 16(1), 294 (1969).

10. M. Kryszewski et al., Disc. Farady Soc. 51, 144 (1971).

11. Sturner and D. Pai, Macromolecules 12(1), 1 (1979).

12. G. Manecke and J. Kautz, Makro Chemie 172 (1), 1 (1973).

13. R. Liepins et al., Polym. Prepr. 11(2), 1948 (1970).

14. R. Liepins et al., Polym. Prep. 9(1), 765 (1968).

15. L. Sonnabend, U. S. Patent 3,700,493.

16. S. Suzuki et al., Jap. Pat. Appl. 7,279,278; CA 12877 (1974).

17. H. Naarmann, Ger. Offen. 1,953,899 (1971).

18. H. Naarmann et al., Ger. Offen. 1,953,898, (1971); CA 37498 (1971).

19. Dutch Patent 6510986 (1966).

20. Y. Fainshtein et al., Vysokomol. Socdin A18(3), 580 (1976).

21. L. Nuttal and W. Tiherington, ASME Procedures 74-ENAS-2.

22. A. Sharpe and R. Windhager, Coatings Conference Proceedings 83 (1981).

23. R. Buvet et al., Ion Exchange and Solvent Extraction 4, 181 (1973).

24. R. bivet et al., J. Chem. Phys. (France) 68, 39 (1971). K. Kanazawa et al., Synth. Metals 1, 329 (1979/1980).

25. D. Wohrle, Makro Chemie 175, 175 (1974).

26. H. Inoue et al., J. Chem. Soc. Japan 65, 238 (1962).

27. W. Slough, Trans Farady Soc. 58, 2360 (1962).

28. H. Pohl and R. Chartoff, J. Poly. Sci. A2, 2787 (1964).

29. A. Dulov et al., C. R. Acad. Sci. USSR 143, 1355 (1962).

30. H. Poly, "Semiconducting Polymers," in Modern Aspects of the Vitreous State, J. Mackenzie, Ed. (Buterworths, London, 1961), p. 72.

31. A. Terenter et al., CR Acad. Sci. USSR 140, 1092 (1961).

32. V. Garten and D. Weiss, Rev. Pur. Appl. Chem. 7, 69 (1956).

33. A. Dulov et al., Izv Akad Nauk SSSR Ser. Khim 909 (1964). A. Dulov et al., Zh. Fiz. Khim. 39, 1590 (1965).

34. R. McNeil and D. Weiss, Aust. J. Chem. 12, 643 (1959).

35. A. Dulov et al., Kokd. Akad. Nauk SSSR 143, 1355 (1962). A. D. Carlton et al., J. Phys. Chem. 68, 2661 (1964).

36. J. Danhaeuser, Makro Chemie 84, 238 (1965). B. Davydov et al., Trans. Poly. Sci. USSR 4, 970 (1963).

Appendix G
Polymer Structures

Typical molecular structures for the polymeric materials included in Appendix E are compiled here to illustrate the structural similarities of different materials that are insulating, semiconducting, or metallic when doped with strong electron acceptors.

I   Poly[N,N'-(p,p'-oxidiphenylene)pyromellitimide]

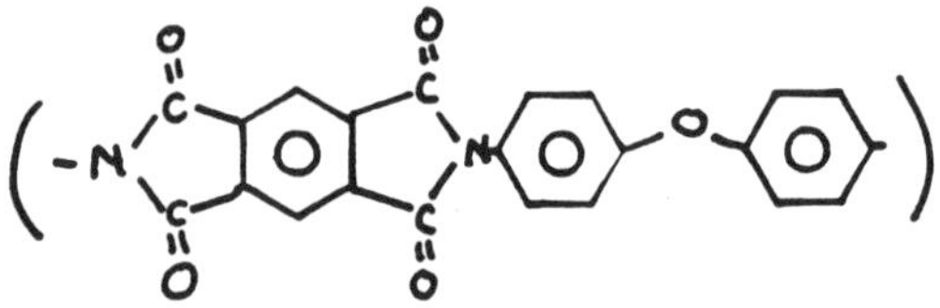

II

   TFE   poly tetrafluoroethylene

   FEP   poly fluorinated ethylene
        propylene copolymer

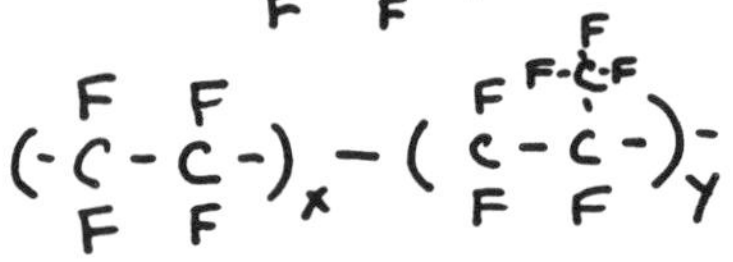

   PFA   poly fluoroalkoxy ethylene

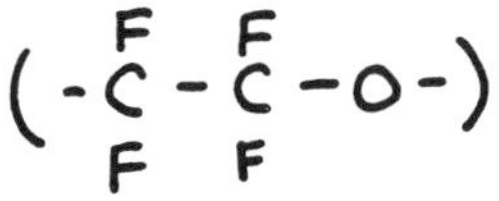

   Kel-F   poly chlorotrifluoro
       ethylene

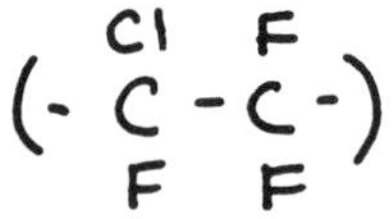

III   poly ethylene terephthalate

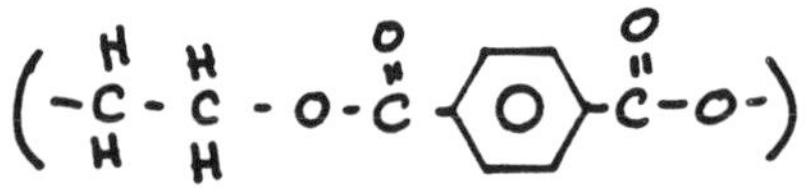

IV

       polyethylene

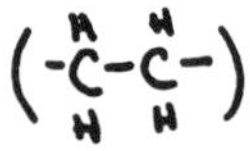

       polypropylene

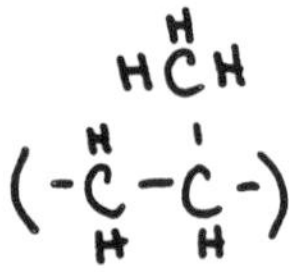

V 1,2,3   Polyacetylene

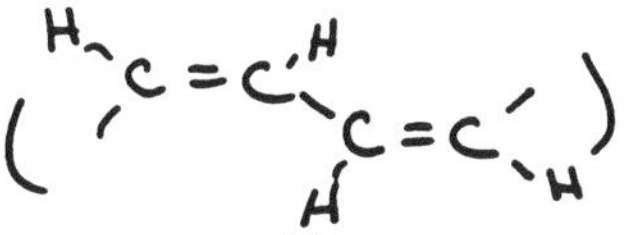

Cis 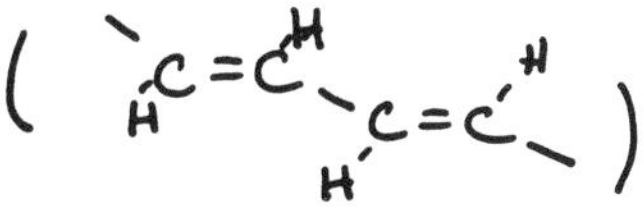   Trans

V5

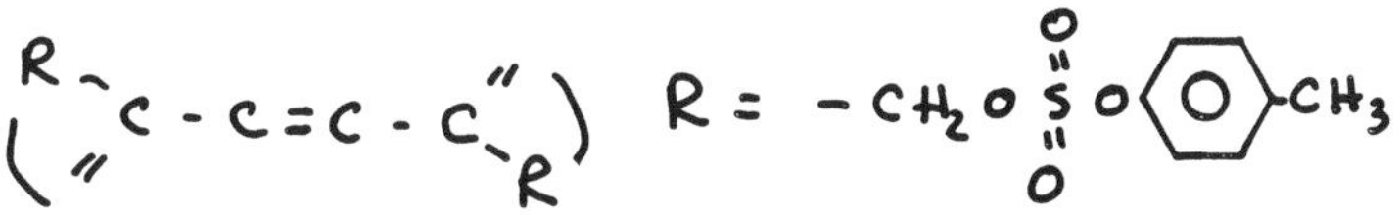

$R = -CH_2 O \overset{\overset{O}{\|}}{\underset{\underset{O}{\|}}{S}} O$ ⟨O⟩ CH_3

Poly(2,4 hexadiin-1,6-diol-bistoluene sulfonic acid)

V4, V6

a,i polyphenylacetylene

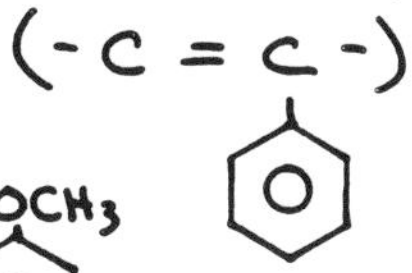

b,j,k,l,m poly(p-methoxyphenyl-
           acetylene)

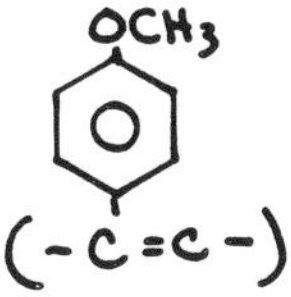

c poly(p-chloro phenylacetylene)

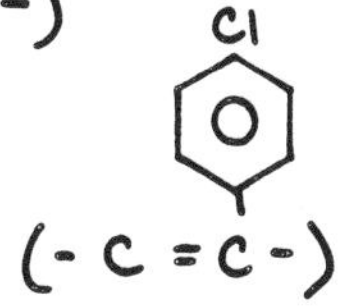

d,n poly(cyano acetylene)

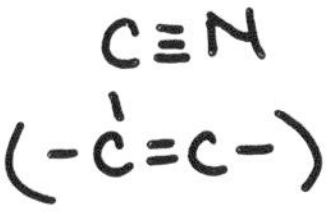

e poly(phenoxyacetylene)

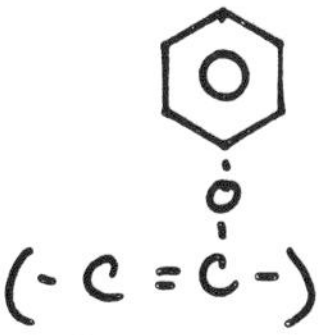

f poly(p-methoxy phenacety-
   lene co pchlorophenyl-
   acetylene) 1:1 role ratio

g 1:2 mole ratio f

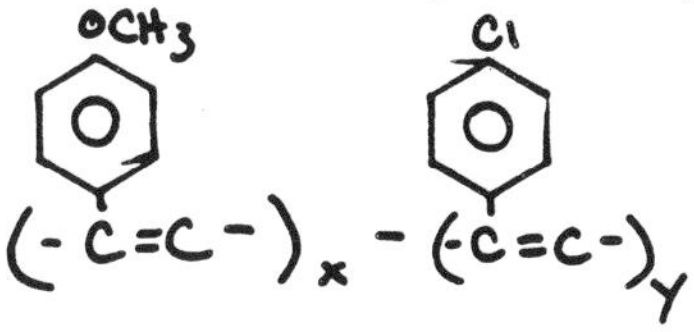

h poly(cyanoacetylene co
   phenoxyacetylene) 1:1
   role ratio

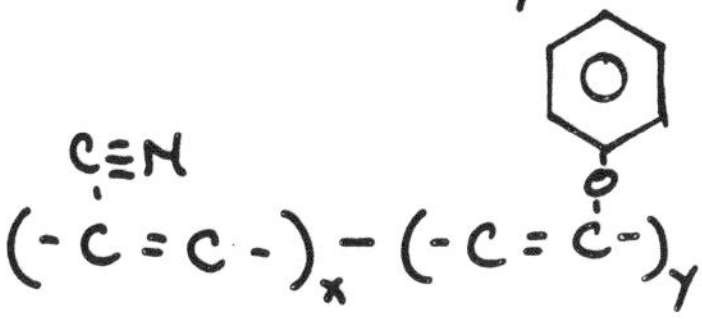

V7   poly(p-phenylenevinylene)

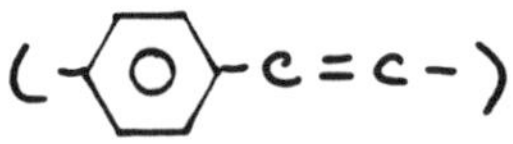

$$(-\langle\bigcirc\rangle-C=C-)$$

V8

  a.   poly(p-amino phenylacetylene)

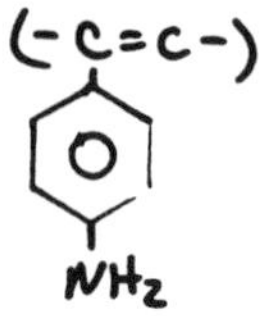

$$(-C=C-)$$

with benzene ring bearing NH_2

  b.   poly(β-chlorophenylacetylene)

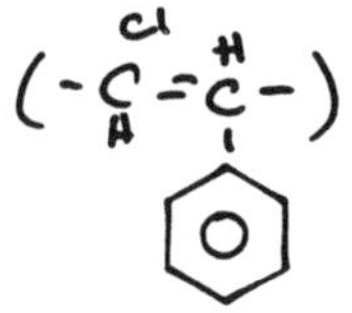

$$(-\underset{H}{C}=\overset{H}{C}-)$$

with Cl and benzene ring

V9

  a.   polypropyne

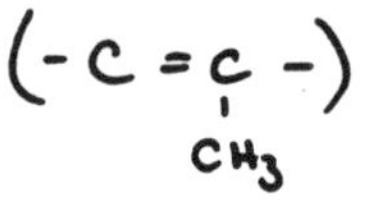

$$(-C=\underset{CH_3}{C}-)$$

  b.   polybutyne

$$(-\underset{CH_3}{C}=\overset{CH_3}{C}-)$$

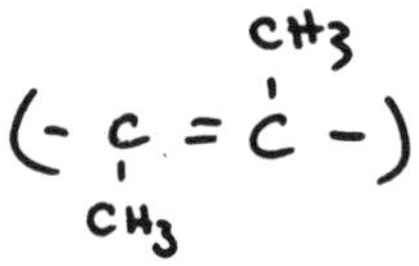

V10   poly(ethynylnapthalene)

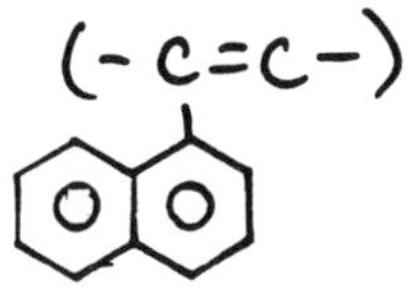

$$(-C=C-)$$

with naphthalene

V11   poly 2,4-hexadyne

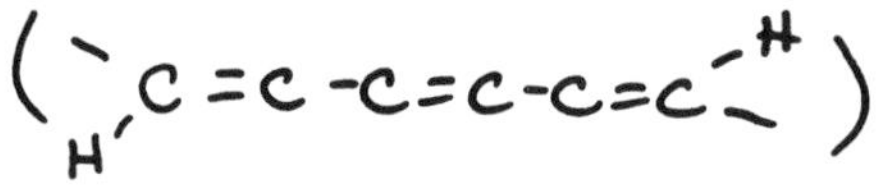

$$(\underset{H}{\overset{\cdot}{}}C=C-C=C-C=C\overset{H}{\underset{\cdot}{}})$$

V12   poly 1,6-heptadyne

$$(-\langle\text{ring}\rangle)$$

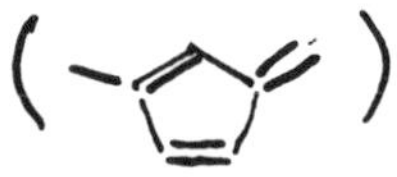

V13   poly tetracyanoethylene

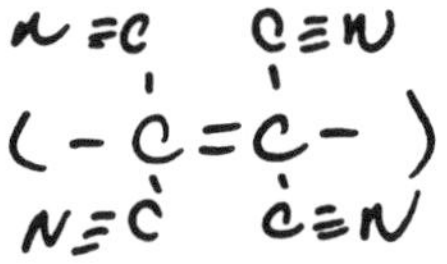

$$\begin{array}{cc} N=C & C\equiv N \\ (-C=C-) \\ N=C & C\equiv N \end{array}$$

VI 1,2,3   poly sulfurnitride (SN)$_x$

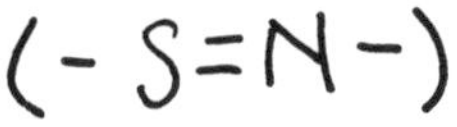

$$(-S=N-)$$

VII   Typical structure,
      Polyacenequinones, branching likely

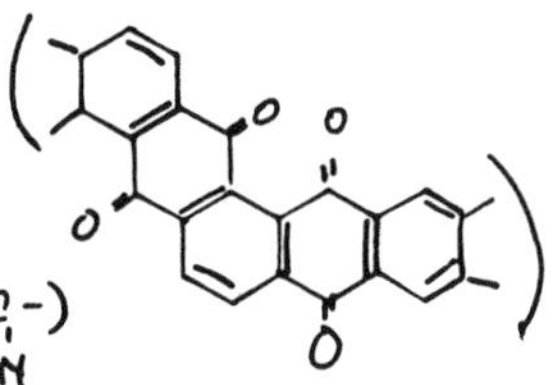

VIII Polyvinyl carbazole

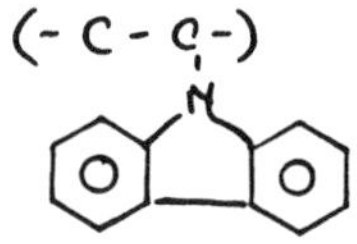

$$(- C - C -)$$

IX   Polyvinylpyridine

$$(- C - C -)$$

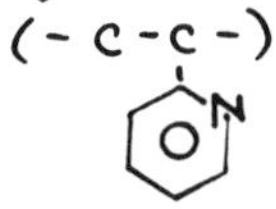

X   Typical structure,
      crosslinked poly-
      phthalocyanine

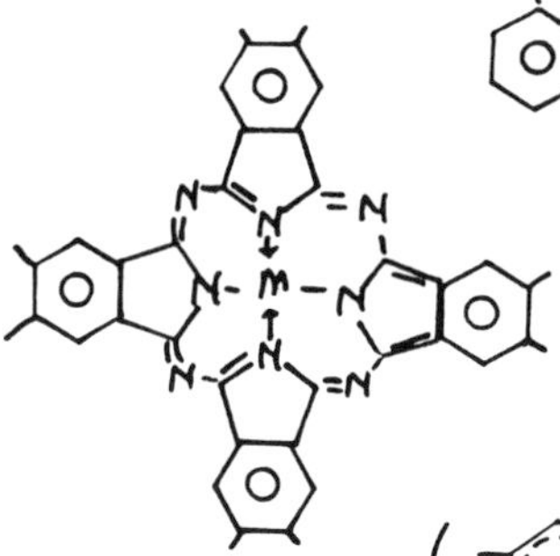

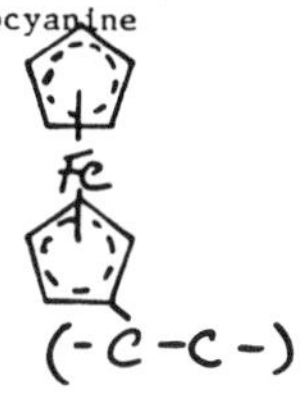

XII-1-a   poly vinyl ferrocene

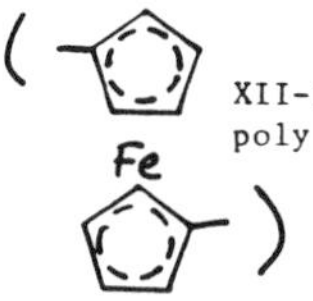

$$(- C - C -)$$

XII-1-b
poly vinyl ferrocenylene

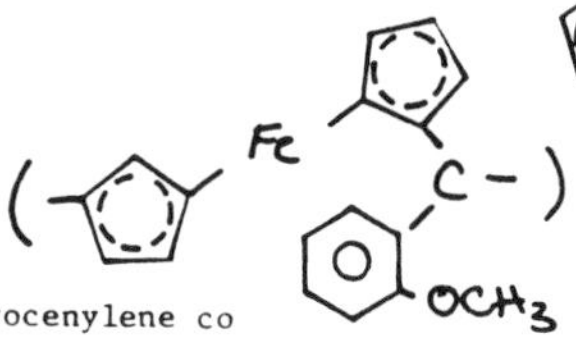

XII-1-c   poly(vinyl ferrocenylene co
          anisaldehyde)

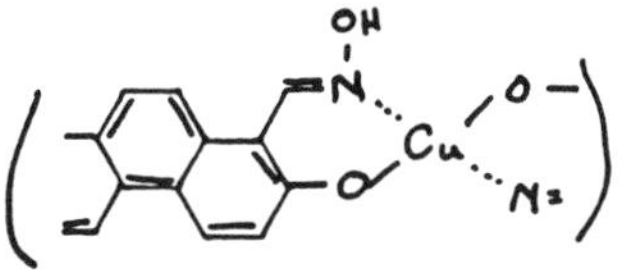

XII-4   poly(1,5 diformyl-2,6-dihydroxynapthalene-
        Cu^{+2}-dioxime)

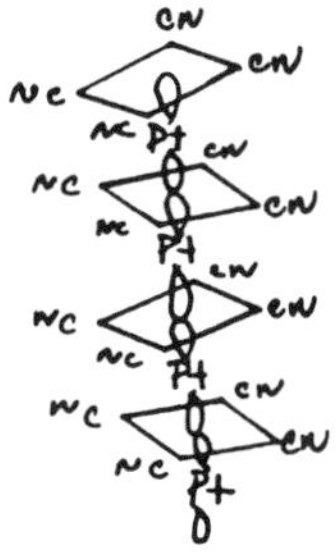

XII-2   [K$_2$P + (CN)$_4$]$_x$

XII 7

Copper–Rubeanic acid

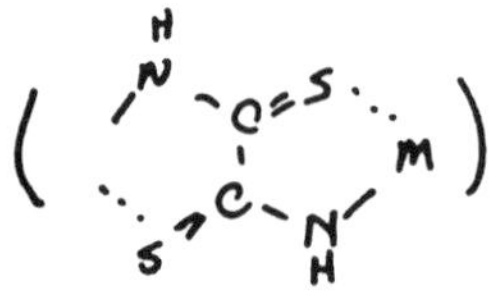

XII 8

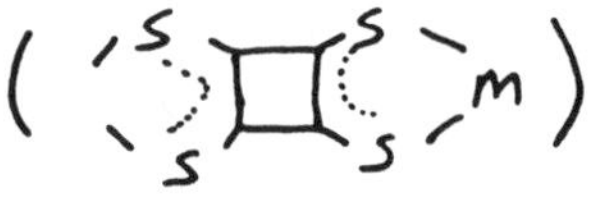

XIII-3 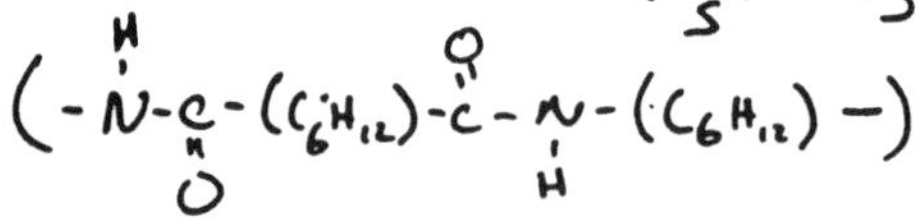   Nylon 66

XIV 1   poly(benzobisthiazole)

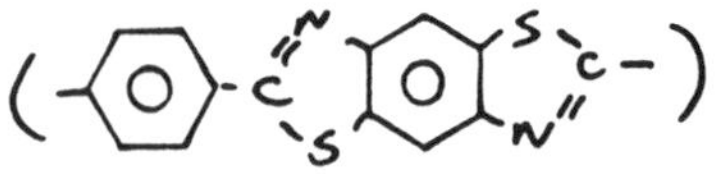

3

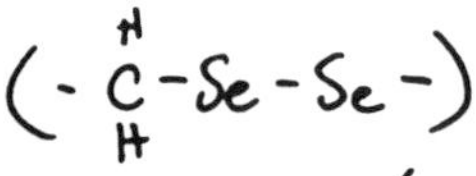

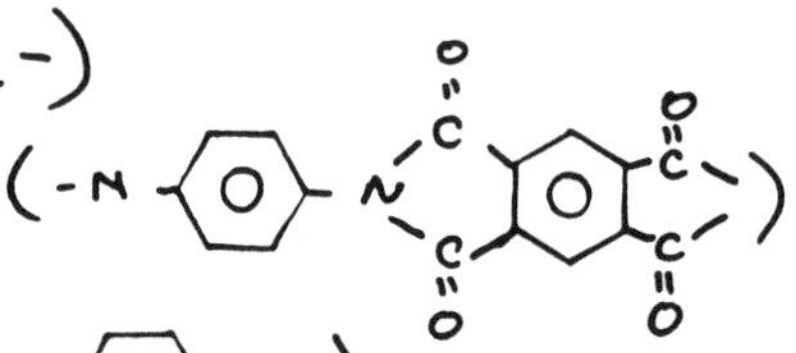

5a

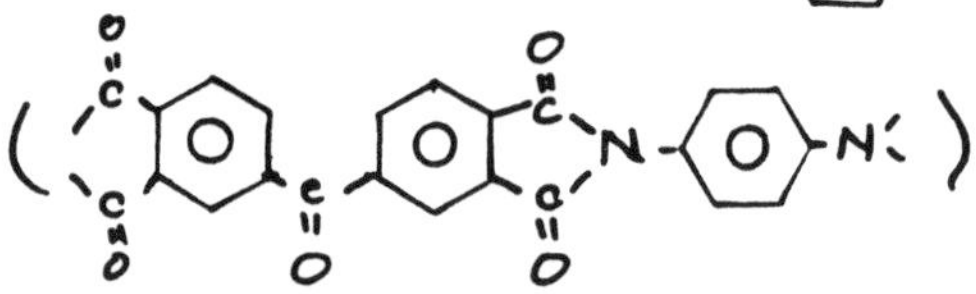

5b

6   emeraldine

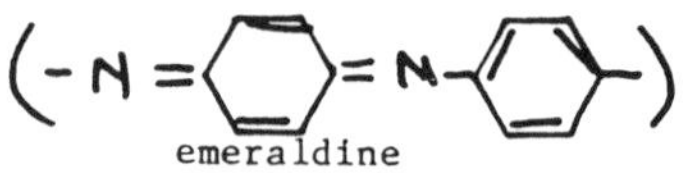

8   polyphenylene sulfide

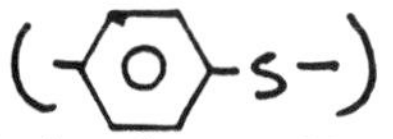

9a

polydimethyl amino styrene

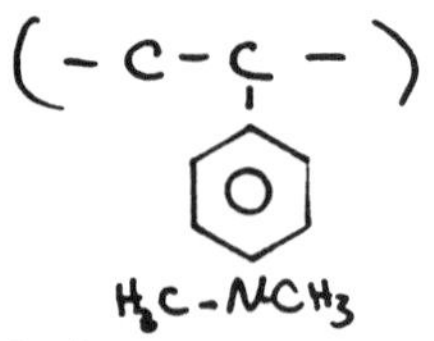

H_3C-NCH_3

9b

poly(vinyl-2-phenothiazine)

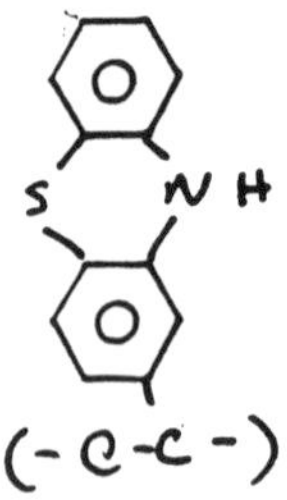

10   Poly(phenyleneamino chloranils)

a

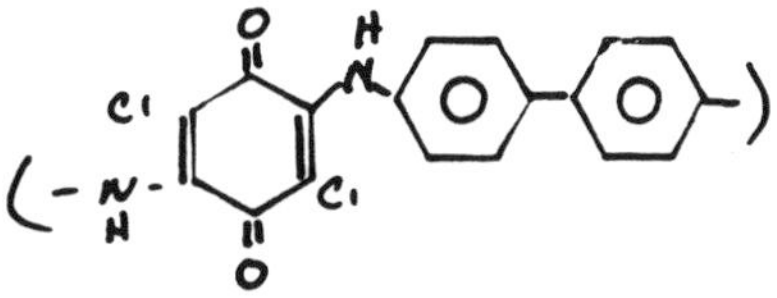

d

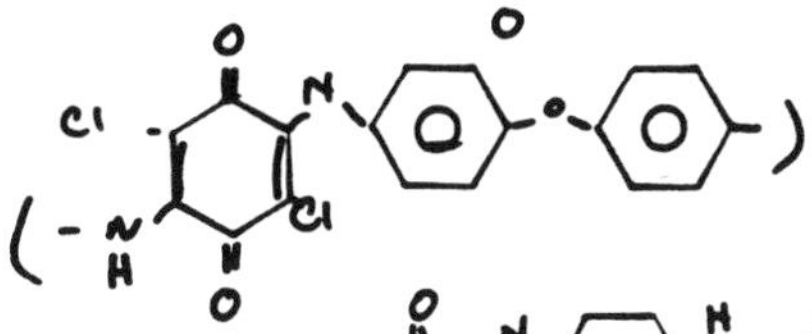

e

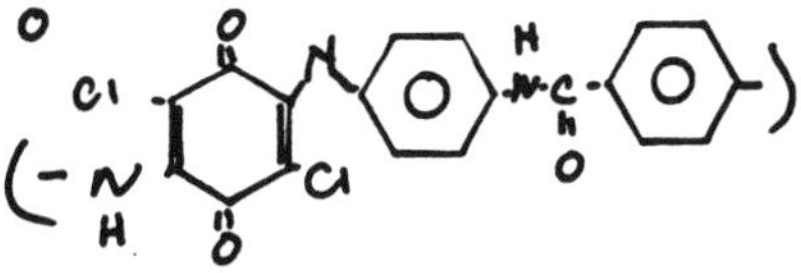

f

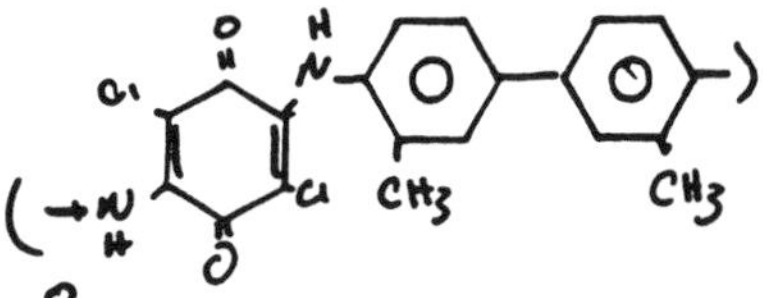

g

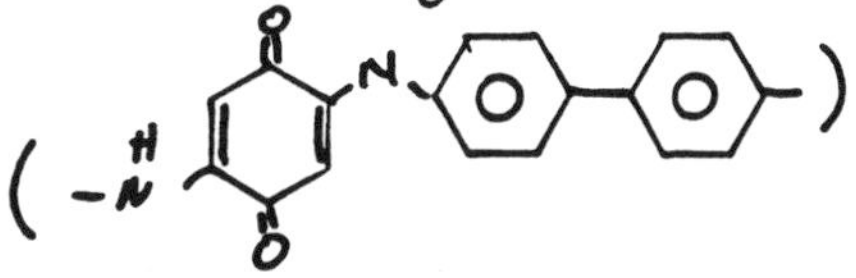

h

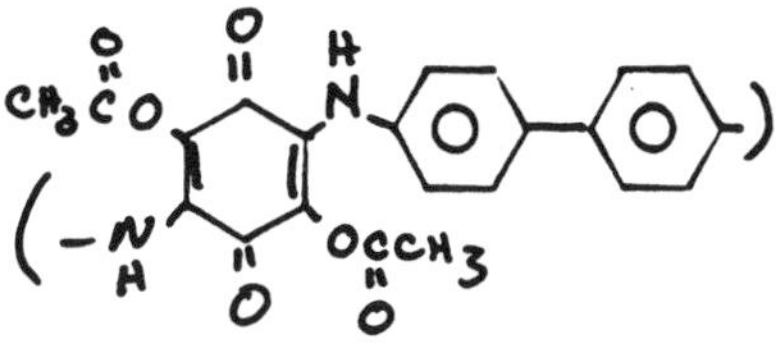

11   Poly(N-carbazolyl-
       ethyl vinyl ether)

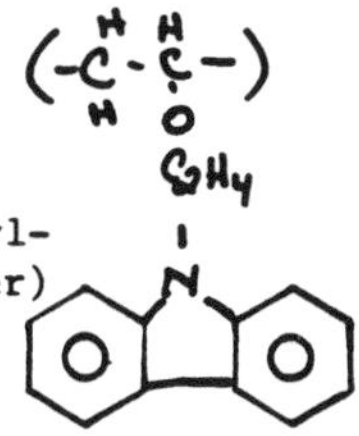

12   Poly(azines)   a

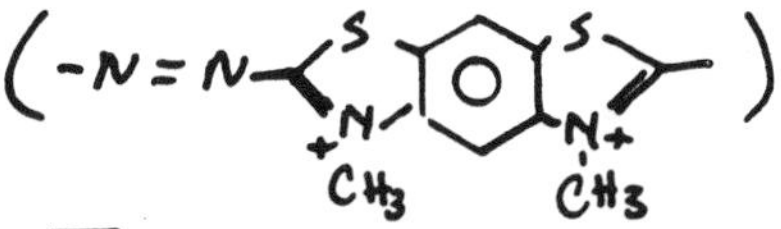

                          b

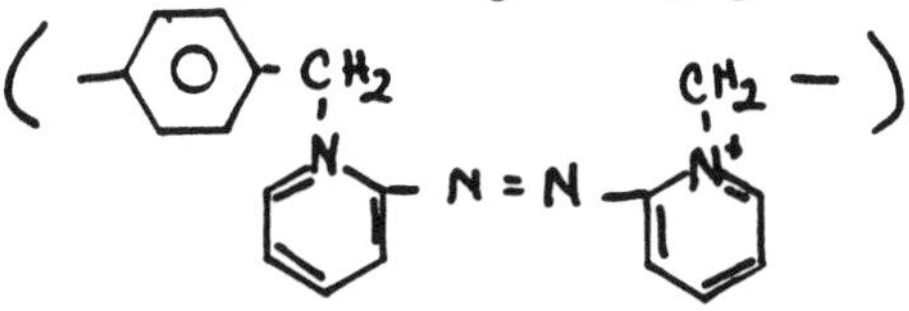

13, 14   1,2 dinitrile polymers

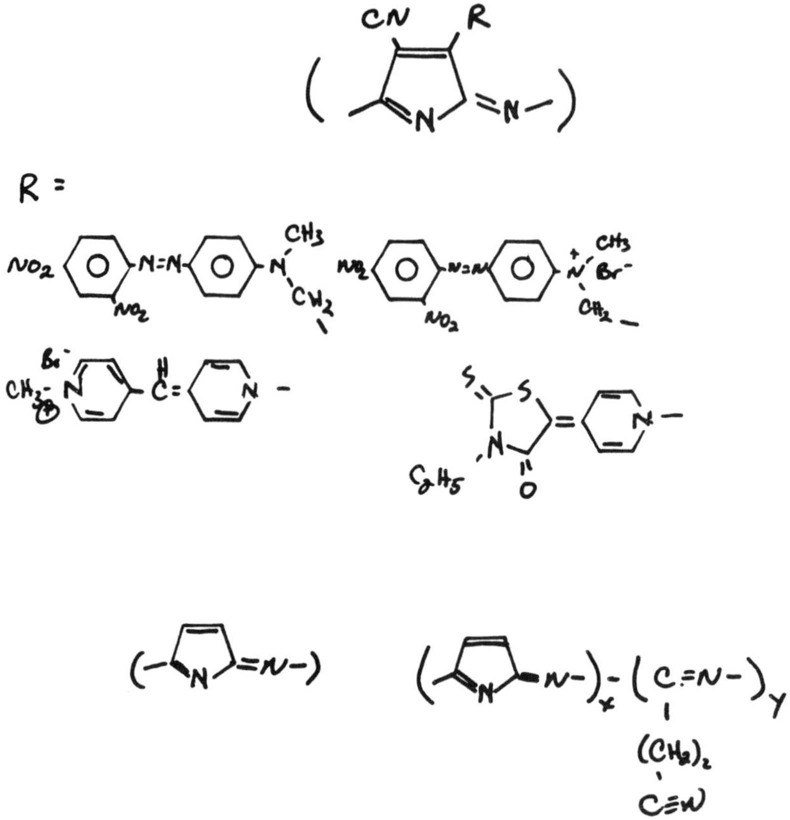

$$\left(\underset{N}{\overset{CN \quad R}{\diagup\!\!\!\diagdown}}=N-\right)$$

R =

NO_2—⬡—N=N—⬡—N(CH₃)(CH₂—) NH_2—⬡(NO_2)—N=N—⬡—N⁺(CH₃)(CH₂—) Br⁻

CH₃—N⁺(Br⁻)=⬡—C(H)=⬡—N—

(thiazolidinone structure) =⬡—N—
C₂H₅   O

$$(-\underset{N}{\diagup\!\!\!\diagdown}=N-)$$ $$\left(\underset{N}{\diagup\!\!\!\diagdown}N-\right)_x\left(\underset{\underset{C\equiv N}{(CH_2)_z}}{C}=N-\right)_y$$

17

(bis-imidazoline structure)
—N—(imidazoline)—N—⬡—
—N—(imidazoline)—N—⬡—

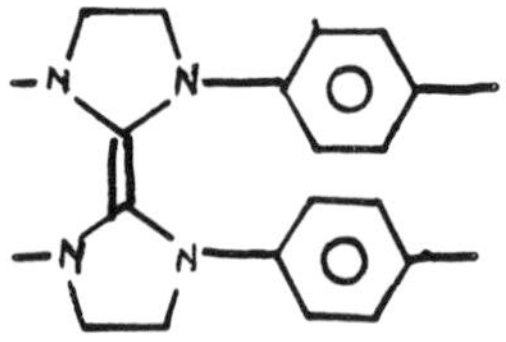

20 Aromatic polyimides
    general structure

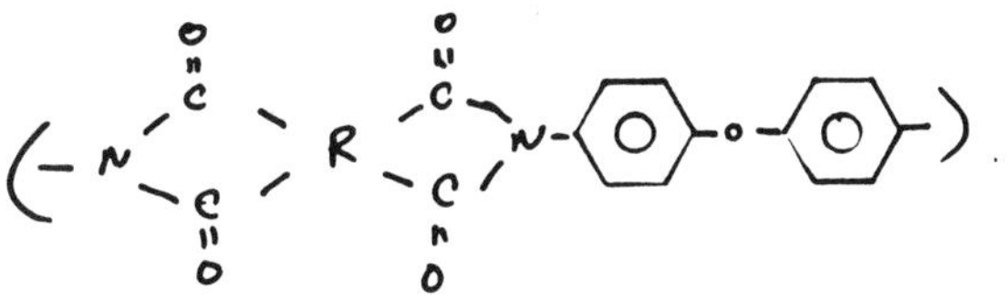

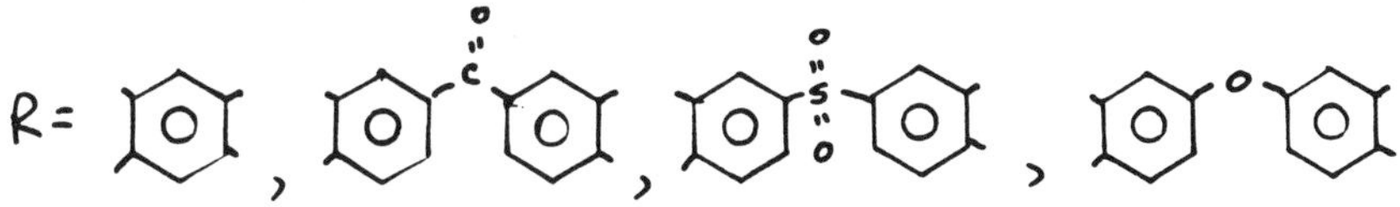

R=

23 Aniline
    black

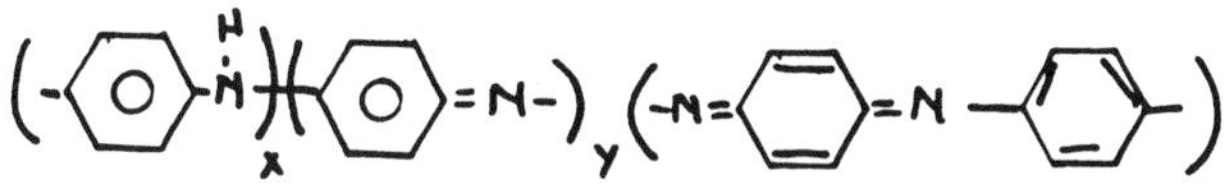

24 Polypyrrole

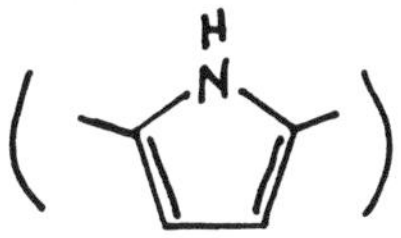

25 polymethinimine

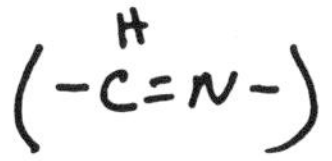

30 poly(tetrachloro phenyl thioether)

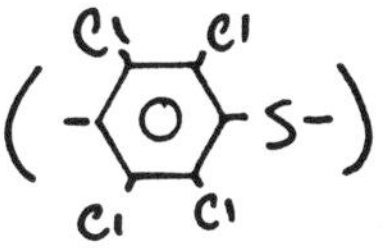

28

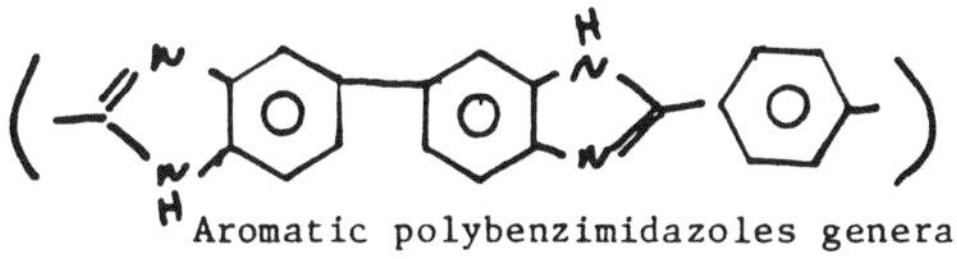

Aromatic polybenzimidazoles general structure

33 poly(arylene
    quinones) general
    wtructure

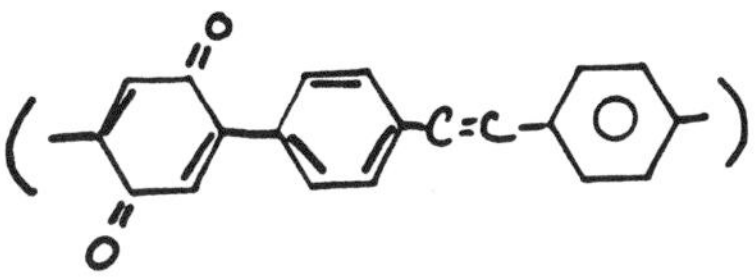

35

Poly(azophenylenes) general structure

Poly(Schiff base) polymers general structure

36